STRENGTH AND DEFORMATION IN NONUNIFORM TEMPERATURE FIELDS

PROCHNOST' I DEFORMATSIYA
V NERAVNOMERNYKH TEMPERATURNYKH POLYAKH

ПРОЧНОСТЬ И ДЕФОРМАЦИЯ В НЕРАВНОМЕРНЫХ ТЕМПЕРАТУРНЫХ ПОЛЯХ

Strength and Deformation in Nonuniform Temperature Fields

A collection of scientific papers
edited by
Prof. Ya. B. Fridman

Authorized Translation from the Russian

Springer Science+Business Media, LLC
1964

The original Russian text was published by Atomizdat, the State Press for Literature in the Field of Atomic Science and Technology, in Moscow in 1962, for the Moscow Engineering Physics Institute.

Library of Congress Catalog Card Number 63-17641
ISBN 978-1-4757-6606-6   ISBN 978-1-4757-6604-2 (eBook)
DOI 10.1007/978-1-4757-6604-2

© 1964 Springer Science+Business Media New York
Originally published by Consultants Bureau Enterprises, Inc. in 1964.
Reprint of the original edition 1964

PREFACE

The often repeated assertion that there is not enough work being done on many problems of mechanical strength is all the more true of thermal strength, i.e., the resistance to flow, creep, simple static, impact and fatigue failure, and the loss of strength brought about by the thermal effects which often accompany mechanical stresses.

There is no question of the great importance of these problems. Thus it is to be expected that the publication of the present collection will be of interest to investigators in various branches of engineering.

In spite of the rather large number of recent papers dealing with different problems in thermal strength, there are only a few comprehensive reviews.*

For this reason, a considerable part of the present collection is devoted to a critical review of the results found in the literature. The presentation of the various papers in the collection is such that they can be used independently. The Strength of Materials Department of the Moscow Engineering Physics Institute extends thanks in advance to all the organizations and individuals who will take the trouble to communicate (address: Strength of Materials Department, Moscow Engineering Physics Institute, Kirov Street, 21, Moscow) their suggestions and critical comments on the material published in the present collection.

---

*Among such reviews we can mention the book by B. Gatewood, and the papers by S. V. Serensen and P. I. Kotov, A. Freudenthal, Manson, et al.

CONTENTS

SOME OF THE LAWS GOVERNING
MECHANICAL AND THERMAL STRENGTH

Ya. B. Fridman

Until recently, effects of mechanical loads (such as weight), inertia, the pressure of solids, liquids, gases, etc., have received most of the attention in studies of strength of materials and structural elements.

At the present time, there are many branches of science and engineering in which very high operating temperatures are used. At the same time the parts required have become more complex. The nonuniform temperature fields that are encountered on going from one part of a body to another (at the same instant of time) and in every part of the body (as a function of time) can often lead to failures, with very small loadings and even with no mechanical loads at all.

In the broad sense of the word, failures of mechanical or thermal strength may be considered to include the following:

1. Reduction of the bearing strength of a structural element (for example, in failure), loss of stability, etc.

2. Failure of the element to operate normally (even while retaining its strength) as a result of excessive deformation, or the development of a failure (for example, deformation beyond the permissible limit), or loss of air-tightness when flaws occur, etc.

In this sense, one may speak of thermal strength or of analyzing thermal stresses, i.e., flow, creep, simple static, fatigue and impact failure, and loss of strength resulting from purely thermal effects. There are various physical fields which lead to nonuniform deformations, which could serve as a basis for treatment of other types of strength — e.g., magnetic or electrical.

Since, in almost all solids the temperature effect produces thermal expansion (or compression), which in the majority of practical cases is restrained,* it is obvious that the stresses and deformations from temperature effects differ from those due to external loads only in the nature of the source.

Since thermal stresses are always determined by the deformation and not by the stresses, they relax to some extent with increase in deformation (or motion), while mechanical loadings may be either strongly relaxing or completely nonrelaxing, for example, in a suspended load (Fig. 1).

A comparison of mechanical and thermal effects is shown for several cases of loss of strength in Table 1.

If it is assumed that an elastic system under thermal stresses is conservative, and that the forces are equal to the derivatives of the potential energy, then for any temperature field, certain surface and volume forces may be calculated the effects of which are equivalent to the effect of the given thermal field. In other words, if a given thermal field is acting, loss of strength can occur at certain critical temperatures, and the critical values of the external loads may be calculated for any given state. Here, a uniform temperature field corresponds to

---

* Thermal expansion of solids is prevented by adjacent zones at another temperature, adjacent parts, or (even in a uniform temperature field) nonuniform or anisotropic properties at various points in the solid.

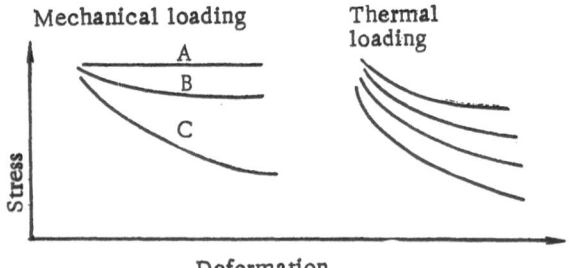

Fig. 1. Relaxation of mechanical and thermal stresses: (A) nonrelaxing mechanical loading (very large elastic energy reserve, case of fixed stress); (B) weakly relaxing; (C) strongly relaxing mechanical loading (elastic energy supply very small). Thermal stresses are, as a rule, strongly relaxing.

the effect of surface loads alone, while a nonuniform temperature field corresponds to the effect of both surface (P_s) and volume (P_v) loads. In this sense, an analogy may be drawn between purely thermal and purely mechanical phenomena. At the same time, however, there are fundamental differences between the effects of mechanical and thermal loadings.

Complete failure (separation) of the part seldom occurs under purely thermal loading. Typical of thermal failure (particularly under multiply repeated loadings) is "saturation" of the part with flaws, which is accounted for by both the local nature and the rapid relaxation of the thermal stresses. Thus, the development of flaws (which under mechanical loading with a given force would increase the stresses and hence accelerate the process), with even small amounts of motion, lowers the thermal stresses and keeps the flaws from propagating, so that they do not get all the way through the cross section of the part. If a thermal load is applied again, the largest thermal stresses occur at other places (since the flaws that have already been formed reduce the amount of local coupling, making local thermal deformation easier, and, thus, "unload" the zones in the immediate vicinity). Thus, whenever thermal loadings are repeated, the flaws are always formed in some new region of the part.

In the present paper,* a short discussion and comparison is given of some of the laws governing mechanical and thermal strength. The discussion is given mainly for the laws governing macroscopic processes, and these may be substantially different from those holding microscopically. For example, microscopic deformation and failure are often discontinuous in nature, although they usually seem to be continuous on a macroscopic scale.

Various Cases of Loading

In view of the great variety of loadings that occur in tests and the even greater variety encountered under operating conditions, we shall consider the possible classifications of the fundamental types of loading. First of all we must distinguish between unsustained and sustained systems.†

In the first case, the loaded part (sample or assembly) transmits from a source with a definite supply of energy, some given initial force or displacement, and then plastic deformation or failure develops with time. Examples of this among mechanical loadings are (1) a bolt, after it has been drawn up, if no additional forces are applied to it, and (2) a tank after it has been given a constant internal liquid or gas pressure. Among thermal loadings, it can be seen in the production of a sustained (steady-state) temperature difference, constant in time, for example, between the inside and outside surfaces of a tube or vessel.

In sustained systems, the part receives the load P or the displacement Δ, not simply at the start, but during the rest of the deformation, either by increase in P or Δ, or by continuous repetition. Examples of sustained systems are provided by the various cases of mechanical and thermal fatigue, as well as by static tests under increase in load (or increase in temperature gradient). Both sustained and unsustained systems can show all four of the kinetic periods of deformation and failure given below. Of course, in sustained systems, the course of the process is determined by both the initial load and resistance, and by the nature of the energy "feed" under load. The distinction between sustained and unsustained systems may change in the course of loading. For example, if the end of failure is passed through very quickly, the sample may become disconnected from the source of loading, and become unsustained. On the other hand, in an initially unsustained system, for example, after

*This paper makes use of the results of work done by the author in cooperation with T. K. Zilova, with B. A. Drozdovskii [1,2] and others, as well as of a number of the papers cited in the references.
†Unsustained and sustained systems correspond to free and forced oscillations, respectively.

TABLE 1. Various Cases of Loss of Strength Under Mechanical and Thermal Loading

	Thermal loading		Mechanical loading without heating	
	Configuration	Critical temp.	Configuration	Critical force
Rod	Uniform heating	$t_{cr} = \dfrac{4\pi^2}{\alpha \lambda^2}$	$N = P_n$	$N_{cr} = \dfrac{\pi^2 EF}{\lambda^2}$
Flat plate	Nonuniform symmetric temperature field; $t = t_0 + t_1\left(1 - \dfrac{r}{R}\right)^n$	$t_{cr} = \dfrac{k}{\alpha}\left(\dfrac{h}{R}\right)^2$, $k = k(n)$		$\left(\dfrac{P_n}{P_0}\right)_{cr} = C_{cr}$, $C_{cr} = f(E,h,R)$
Curved plate	Uniform heating	$t_{cr} = f(\alpha, E, h, L, m, R)$		$P_{ncr} = f(E,h,R,L,m)$
Cylindrical shell	Nonuniform temp. field; $t = t_0 f(x)$	$(t_0)_{cr} = f(\alpha, E, h, R, L)$	$N(x) = N_x f(x)$	$(N_0)_{cr} = f(E,h,R,L)$

formation of a neckdown or a flaw, the system may become sustained by a narrow region being continuously fed with energy from other parts of the body as it becomes isolated.

Another important distinction may be drawn between methods of loading, on the basis of the way in which the load is applied to the part: there may be a fixed load, i.e., one which does not relax during deformation, or the displacement may be fixed, or there may be mixed forms. Examples are given in Table 2 from which it may be seen that it is typical of temperature loading to have fixed displacement (since what the temperature gradient determines is the displacement). The importance of this fact is shown in the paper by N. D. Sobolev and V. I. Egorov, "Thermal fatigue and thermal shock" (see the present collection, page 62).

Four basic cases should be distinguished for either thermal or mechanical effects, depending on the nature of the loading:

1. Mechanical or thermal shock, in which wave processes begin to play a substantial role (e.g., in sudden local bathing of a part of considerable length by metallic coolant), as do inertial forces (e.g., even in light metals the effect of a shock velocity of several hundred meters per second is determined mainly by the increase in inertial resistance, rather than by change in mechanical properties). Under static loads, neither wave nor inertial corrections play any essential role. This is the specific feature of shock loading which does not show up under static load. Note that many mechanical properties usually designated by the word shock (or impact), are not of this nature at all — e.g., the "impact toughness" when a notched sample is bent with a swinging pendulum

TABLE 2. Comparison of the Effects of Various Factors on Temperature Stresses of the First and Second Kinds [10]

Type of factor	Special features of macro- and microtemperature stresses	
	Macrotemperature stresses (of the first kind) occur, as a result of temperature difference between different zones in the solid. The surface zones are usually more strongly stressed than the interior zones, and thus the dimensions of the surface zones are less than those of the interior zones. As a result, in rapid heating, the total deformation of the solid is considerably less (the external hot zones are small), than in rapid cooling (the extent of the interior hot zones is large).	Microtemperature stresses (of the second kind) occur, especially when the lattice has low symmetry, for example, in cadmium, tin, etc., as a result of anisotropy and difference in thermal expansion of the elements in the structure. Relaxation of microstresses, because of their large extent and dispersion, goes comparatively slowly when kept at elevated temperatures.
Kept at the upper temperature of the cycle	Little effect, since (with the exception of a steady-state temperature gradient) stresses only occur when the temperature is changed.	Large effect, since the structural inhomogeneity which produces the microstresses continues to act even when the solid is at a constant temperature.
Change in heating and cooling rates	Large effect, determining the magnitude and sometimes the sign of the stresses.	Small effect, since microrelaxation processes usually cannot occur to any appreciable extent when the temperature is changed.
Shape and absolute dimensions of the solid	Small effect, particularly with increase in the surface-to-volume ratio (for the same value of cross section).	No appreciable effect, since the microstresses are determined mainly by local conditions.
Grain size	Little effect. Often no difference between single and polycrystals.	Increasing the grain size reduces the number of structural elements, and thus greatly reduces the residual deformation when the temperature change is repeated.
Texture	Small effect.	Large effect, sometimes changes the growth coefficient in temperature changes by a factor of 20.

4

striker. It is for this reason that almost all the properties revealed by impact toughness tests may also be found from static tests of notched samples, and so it has been proposed to use the words notch toughness instead of shock (impact) toughness [3].

2. Short-time static loading, in which the growth time of the static mechanical or thermal load (usually of the order of minutes) is commensurate with the total duration of the loading. Some examples are the usual static mechanical tests (for elongation, hardness, etc.), or a deformation produced gradually by slow change in temperature.

3. Long-time static loading, in which the growth time of the load is usually small in comparison with the duration of loading, which is measured in months or even years. The most dangerous of the long-time loadings is pneumatic loading, especially from large pressure sources [the load (see Fig. 1A) is almost unrelaxed with increase in deformation]; less dangerous is hydraulic loading, since the load relaxes weakly (see Fig. 1B); and even less dangerous is long-time loading with a fixed deformation [the load (see Fig. 1C) relaxes strongly]. An example of long-time thermal loading is creep under thermal stresses, which may lead to failure after the "critical time" has been reached.

4. Repeated static loading produced by going through mechanical or temperature cycles (from hundreds and thousands to millions of times), which, respectively, produce mechanical or thermal fatigue. In the latter case, we are usually speaking of temperature cycles with large amplitudes, of the order of hundreds of degrees. The maximum total number of such temperature cycles under operating conditions is often considerably less (from 10^2 to 10^3) than that of mechanical cycles (which, in turbine blading, for example, can get as high as many millions). Some of the parts of physical energy installations have an exceedingly small load repetition frequency, e.g., one cycle per week or even one per month. Under these conditions, the effect of the long-time strength and relaxation characteristics is very important, especially at high temperatures.

Thus, of the four cases of loading mentioned, the last three — short-time, long-time, and repeated — are static, and the third — long-time — usually occurs in initially unsustained systems, while the rest occur in sustained systems.

The Three Regions of the Deformation Process

When sufficiently large forces or displacements (acting for long times) are applied to a part, three basic ranges (Fig. 2) are found to be successively superimposed on one another during mechanical or thermal deformation:

1. The elastic range, which disappears when the load is removed. Pure elastic deformation is very seldom encountered under operating conditions, especially if there is heating. Usually, immediately after the elastic range, one or both of the following stages occurs to some extent.

2. The plastic range, or better, elastoplastic range, since the elastic deformation, proportional to the applied stresses, is retained even after the plastic stage occurs.

3. The failure range, which is almost always also accompanied by elastic and plastic deformation, the elastic deformation at the end of the process usually being localized near the failure flaws.

Although the boundaries between these three ranges are rather arbitrary (and depend primarily on the local method of observation), it is still necessary to make the distinction. In many plastics, in addition to these three regions, an important role is played by highly elastic deformation, which sometimes reaches hundreds of percent at a very low modulus of elasticity (1 kg/mm² or less).

In each of the three regions, we can distinguish two groups of strength defects [4]: (1) Gradual, as a result of excessive size of the deformation (elastic, E, or plastic, P), or as a result of the excessive number or large dimensions of the flaws (arrows in Fig. 2). In these cases, there is usually no reduction in bearing strength (the load carried by a given assembly), but normal operation may be interfered with as a result of excessive deformation or incipient development of failure flaws. Since the word "excessive" is arbitrary, and depends on the operating requirements of the assembly, gradual losses in strength are usually defined in terms of the "tolerance" on the extent of deformation or flaw formation. (2) Sudden (discontinuous) losses in strength, usually with the load

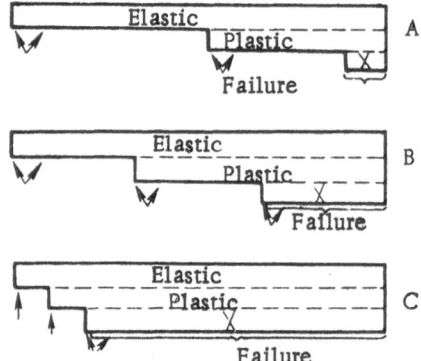

Fig. 2. Diagram showing the relation between the elastic deformation, plastic deformation, and failure regions in samples: (A) smooth uniformly loaded; (B) notched or with irregular structure; (C) real part with large irregularities in resistance and loading. The X shows the critical failure states, and the arrows show the "allowable" deformations.

going through a maximum while reaching a definite value of elastic or plastic deformation, as well as explosive-type mechanical failures* after the amount of deformation or the number of flaws has reached the critical value (for the conditions existing). These losses in strength are shown by the X's in Fig. 2. They include many cases of loss in thermal (elastic or plastic) strength, as well as of mechanical failures (brittle, fatigue, long-time static, etc.).

If the conditions of loading are such that the force is kept constant (for example, by the weight of the load), the discontinuous process can naturally not show up as a change in force. In such cases, what is observed is a sudden increase in the rate of deformation and failure. Note that the so-called "instant of failure," which shows up clearly, for example, in short-time static tests on quite strong and heavy samples in the form of audible and kinetic effects (the testing machine shaking, etc.), represents neither the beginning nor the end of failure, but simply corresponds to a sudden acceleration of the process.

Kinetic Study of Strength and Deformation

The classical theory of strength was developed mainly on the basis of the idea that there is a relation between the stresses (loads) and the deformations (displacements).

The laws on which the theory of elasticity (Hooke's law) and the theory of plasticity (e.g., the generalized Ludwig curve) are based establish a unique relation between the various stress and deformation functions. The time (or rate of the process) is completely absent from these laws, since it is assumed that during a negligibly small period of time (as compared with the total length of time the part is under load) a stable value (for this loading) is set up for the elastic or plastic deformation. Note that in addition to the theory of elastoplastic deformations (where a relation is set up between the differentials of the stresses and the deformations) there is the theory of flow, in which it is shown that the increments to the plastic deformation (rate of plastic deformation) are proportional to the stresses at the particular instant. In other words, the stresses determine the instantaneous increments to the plastic deformation.

In recent years, it has become more and more usual to take account of time and velocity in quasi-static processes as well as in wave processes, i.e., it has become necessary to take a kinetic approach to the problems of strength and deformation. This is due to:

1. The more and more extensive use of high temperatures.

2. Higher mean stress under operating conditions. (Note that sometimes this factor alone is sufficient to make strength and deformation a function of time even at room temperatures, as in the case of "cold creep" and "delayed failure" in structural steels, titanium, and aluminum alloys at 20°C.)

3. Having operating conditions (even at small mean stress) where there are local overloadings and generally nonsteady states.

4. Largely nonuniform and transient character of the stressed state at various places, caused by complex shape and the wide use of assemblies in which materials are used together which have different coefficients of expansion, different heat transmission and heat conduction, different elastic and plastic properties, etc.

---

*Explosive-type failures are often inaccurately called instantaneous. In the present paper, we are talking about purely mechanical "explosions" without chemical energy entering in. The explosive nature here is due to sudden liberation of a large amount of potential and kinetic energy.

5. The wider and wider use of materials of complex composition and structure which, since they are unstable in the physicochemical sense, undergo structural changes with time under load, particularly on the surface, where they are most subject to external effects.

All these five facts above force us to adopt a kinetic treatment which includes the velocities and accelerations of deformation and failure, instead of a quasi-static approach.

Even in the first studies made on creep and long-time strength at elevated temperatures, the dependence of the results obtained on the testing time (velocity) showed up so clearly that from then on these groups of characteristics have been a classic example of the time dependence of mechanical properties, which was subsequently observed at lower temperatures as well. In the half century that has passed since the first studies were carried out, a detailed study has been made of the kinetics* of creep. The classical curve of creep may be divided into three† basic regions or stages:

1. Unstabilized (attenuating, or β-flow)

2. Stabilized, or uniform (quasi-viscous, or K-flow)

3. Accelerated (final).

Although it has been observed in some of the work that the accelerated stage of creep and failure starts at the instant of loading, i.e., there is an initial accelerated or incubation period similar to what occurs in a number of cases of brittle and fatigue failure, the opinion is widespread that "the incubation period does not represent a normal creep phenomenon in polycrystalline aggregates under normal stresses" [6]. Accordingly it has so far been customary in handbooks, text books, monographs, and strength calculations to divide creep into the three stages given above.

Studies of the kinetics of plastic deformation and failure made with a controllable supply of elastic energy lead to the conclusion that having an initial accelerated or incubation period in the creep is neither a special case nor an exception, but represents the same sort of general law as having attenuating or stabilized creep.

The reason the incubation period has not been observed experimentally in most of the work done is probably that the length of the initially accelerated or incubation period (usually fractions of seconds or minutes) was considerably less than the time from the start of loading to when the first deformation reading was taken (fractions of an hour or even several hours). Here, two fundamental cases may evidently be distinguished:

1. The incubation (initially accelerated) period is so short that it is negligibly small in comparison with the total duration of the creep process. In this case, the incubation period may either form a part of the actual loading period, or it may continue somewhat longer. Sometimes, the start of deformation in creep is called instantaneous or sudden deformation. This, however, is incorrect, since the deformation propagates at a finite velocity, less than the elastic wave velocity.

2. The incubation period is commensurate in time with the subsequent stages of creep.

In the vast majority of laboratory tests of creep samples, the incubation period remains hidden. However, this does not mean that the incubation period is of small practical importance, since in some cases of loading (e.g., with the large temperature changes in parts of rockets, reactors, jet engines, etc.) the total time of operation at elevated temperatures is measured in hours, and sometimes even in minutes and seconds.

A study of the incubation period is, in our opinion, of great theoretical importance, as is the case, for example, in the kinetics of chemical reactions or phase transformations.

Note, for example, that in investigating the microkinetics of the martensite transformation [7], use is also made of the derivatives of the amount of martensite formed per unit of time in tension; an induction period,

_____

*The kinetics of microscopic and submicroscopic processes (often of a discontinuous rather than monotonic nature) is not discussed here.

†In some papers [5], a fourth region is distinguished, corresponding to sudden increase in deformation after the flaws reach a critical dimension, but this classification takes no account of the incubation period.

where the process receives positive acceleration, is also observed. Most often, the kinetics of plastic deformation and failure is investigated from curves of "relative deformation ε as a function of time τ" or, much less frequently, from curves of "creep rate V_ε as a function of time."

It is well known that the course of the process is very sensitive to change in velocity, i.e., the acceleration of the process. Here it is important to distinguish between the acceleration j, as used in the mechanics of solids, and the acceleration of the relative deformation j_ε, as calculated for the simplest case of uniaxial deformation.

It is obvious, in particular, that $j_\varepsilon = j/l$, and $V_\varepsilon = V/l$, where l is the linear dimension, i.e., that V_ε and j_ε have the dimensions $\sec^{-1}$ and $\sec^{-2}$, respectively.

It should be noted that a rigorous determination of the rates and accelerations for failure is considerably more complicated than that for deformation, since definite relations have to be set up between the flaw lengths, the relative deformation, and the time.

Placing limitations on the derivatives of a monotonic function is more rigorous than on the functions themselves, and thus, constructing (j_ε, τ) curves gives a more sensitive measure of the process than drawing the ordinary creep curves. In particular, constructing the second derivative from the usual creep curve, starting with the attenuation of the creep, immediately shows that it is necessary to take account of the incubation period. As a matter of fact, the process cannot start with a retardation (i.e., with negative acceleration operating), and it is thus obvious that attenuating β-flow must be preceded (although only for a very short time) by an incubation stage of initially accelerated creep.

The further development of the process depends on the amount of slowing down encountered by the developing deformation or failure. The slowing down may be due, for example, to cold working or structural processes. If the growth rate of the hardening processes is too slow (e.g., under repeated loadings where overloading occurs), or if there is for practical purposes no hardening (e.g., in failure of silicate glasses, quartz, and similar materials), the rate of the process continues to increase, sometimes reaching a limiting value which is maintained up to complete failure. According to the classification given below, this corresponds to direct transition of the first incubation period into a fourth, self-accelerating period.

If, however, as is observed in the majority of creep tests, hardening processes (which increase the resistance of the sample), such as cold working and structural hardening, enter into the deformation during the incubation period, the incubation period is replaced by the next, or slowing down stage (the second period in the classification given below).

What has been said is supported by the creep curves obtained by Tapsell [8], from which it may be seen that the less the hardening, and thus the slowing down of the process (at the temperature at which the creep tests were made), the more clearly the incubation, or initially accelerated, period shows up.

If the hardening is then reduced,* a third kinetic period occurs, consisting of uniform or stabilized K-flow, until, finally, it passes to the final self-accelerating period.

Localization of the deformation in definite zones of the sample (either as a result of structural changes, or in particular as a result of the development of flaws which are observed even in the K-flow stage) produces a change from an unsustained system (which on the macroscopic scale is the system formed by the sample and the machine with the suspended load) to something similar to a sustained system, since the sustaining force comes from the zones in the sample itself adjacent to the local deformation and failure zones. As the local failure zones develop, they receive more energy, and the rate of the macroscopic process increases.

Thus, in the development of flow, creep, and failure with time, we can generally distinguish four periods:

1. The induction, or initially accelerated period — the start of development of the process with an increasing rate of deformation v_ε and of failure $v_{tf} > 0$, and positive accelerations of these processes, j_ε and $j_{tf} > 0$.

---

*We are talking about the resultant effect of hardening and dehardening, which is probably in all cases the resultant of both processes going on at the same time but to different extents.

2. Slowing down — the rate of the process decreases, and the acceleration is negative, which is the result of increasing strengthening of the material as a consequence of cold working or structural processes.

3. The quasi-steady state or quasi-viscous period — the rate of the process is constant, and the acceleration is equal to zero (corresponding to the period of stabilized creep or steady-state development of failure). In special cases, the rate may become equal to zero, which means that deformation and failure have stopped developing (for example, stabilized flaws). As a result of accumulation of microscopic processes (structural changes, development of dislocations, microflaws, etc.), the average resistance of the sample changes, and sometimes so does the force acting on it. Under definite conditions, these changes lead to a transition from the quasi-steady state period to the next, or fourth, period.

4. The self-accelerating (sometimes of avalanche type) final period — where there is a large increase in the degree of localization of deformation and failure, the rate of the process increases, and the acceleration is positive.

The duration of these four periods and transition from one to another are determined by the relation between the rate at which energy is supplied to the loaded body, and the rate at which it is absorbed. The rate at which the loaded body is supplied with energy depends on the external loading conditions, but, after local deformation and failure zones have appeared, it depends also on the supply of elastic energy in the other parts of the body which are supplying the failure zones. The rate at which the energy is absorbed depends on the external conditions, as well as on the shape, dimensions, and structure of the material in the loaded body and, after the localization zones have appeared, it depends on the energy absorption by these zones.

The relation between the loading and absorption characteristics determines the force and deformation kinetic curves in the loading process.

There may be a transition from the first period directly into the fourth. At normal and low temperatures, this is usually observed in brittle failure of metals and glasses, as well as in fatigue failures with overloading. In long-time static strength tests, it occurs when test temperature is high, or the stresses large, or when the hardening processes are only weakly developed.

The similarity between the laws governing the kinetics of plastic deformation and that of failure (clearly visible, for example, from a comparison of the kinetics of creep and that of fatigue failure) may obviously be accounted for by the fact that both processes are based on shear deformations, which prepare the way for subsequent failure, and are superimposed on it. It was shown long ago that all practical cases of failure are accompanied by plastic deformation, although it is sometimes of a very local sort. Note that the opinion sometimes expressed, that flaws "strengthen" the sample does not seem to us to be true. If flaws are present, and the sample gets stronger, it is not because of the flaws but in spite of them. In other words, hardening can occur in a sample (from cold working or structural transformations) even though flaws are present.

The "failure stresses" found from experiments usually do not reflect either the beginning or the end of failure, but only the transition from the precritical (first three) kinetic periods to the postcritical (fourth) period, in which the acceleration of the process is positive and does not decrease. The precritical periods correspond to the idea of developing failure, while passing through the critical state corresponds to regarding failure as a jump.

By analogy with the study of the damage done in fatigue, it has been suggested in [9] to construct curves showing the amount of damage done in evaluating long-time resistance to heat.

Constructing curves showing the acceleration of deformation and failure [for example, by taking the second derivative of the (ε, τ) and (l_{tf}, τ) curves] is a help in distinguishing the fundamental kinetic stages, particularly the incubation period, as well as making it easier to predict the long-time behavior of the material from the results of relatively short-time experiments.

According to the correspondence principle, as developed in present-day physics, the laws governing a more general theoretical approach will reduce to the laws governing a previous, more special theory, as some definite characteristic parameter approaches a limit. For example, as the value of the quantum of action approaches zero, the equations of quantum mechanics approach the equations of classical mechanics asymptotically.

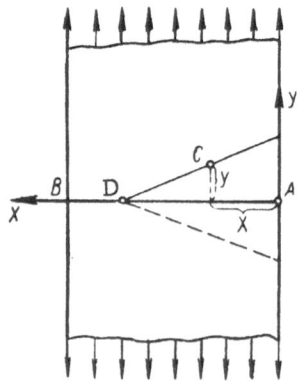

Fig. 3. Development of flaws in a plane sample in tension.

The characteristic parameter for the kinetics of deformation and failure is the time (or rate) of the process.

It is natural to expect that the kinetic approach will also include as a special case the already familiar quasi-static laws of strength and deformation, if the rate of the process approaches infinity (or the time approaches zero). From the development in time of any failure process (or any deformation), it may be shown that for a failure that occurs all at the same time (at the same cross section), for example, by breaking, the velocity of propagation of the flaw must have been infinitely large. Consider the simplest case[*] of a plane sample in tension (Fig. 3), in which a flaw is formed at the point A, moving at the constant velocity v_x. We have $\tau_x = x/v_x$, or $x = v_x \tau_x$, and at $x = 0$, $\tau_x = 0$. At the instant of time $\tau > \tau_x$, the longitudinal path of the point $C(x,y)$ is equal to

$$ y = \left(\tau - \frac{x}{v_x} \right) v_y, \tag{1} $$

and since $y > 0$, the equation has meaning only for $\tau \geq x/v_x$.

It follows from Eq. (1) that

1. At $x = $ const, expression (1) gives the equation of motion of the point with abscissa x along the axis of the sample (y axis);

2. At $\tau = $ const, expression (1) gives the equation of the outline of the flaw at any instant:

$$ y = - \left(\frac{v_y}{v_x} \right) x + \tau v_y. $$

It is obvious that as the ratio v_y/v_x decreases, there is a decrease in the angle of opening of the flaw. Over the entire cross section, the equation is written as y = const for $\tau = $ const, or dy/dx = 0, or

$$ \frac{v_y}{v_x} = 0, \tag{2} $$

but $v_y \neq 0$ (since for $v_y = 0$, the flaw is not opened up at all). Accordingly,

$$ v_x = \infty, \tag{3} $$

i.e., failure occurring all at the same time corresponds with an infinite velocity of propagation of the flaw in the transverse direction.

The same result may also be obtained for the more general case $v_x \neq$ const and $v_y \neq$ const for an arbitrary, not necessarily rectangular flaw. Let the equation of motion of the vertex of the flaw along some line AB (see Fig. 3) be $s = f(\tau)$. The inverse function, $\tau = \psi(s)$, gives the time at which motion starts for a point at a distance s from A. For $s = 0$, $\tau = 0$.

The path of the point C at the instant of time $\tau > \psi(s)$ is given as a function of the time during which the point C takes part in the motion:

$$ y = \varphi[\tau - \psi(s)]. \tag{4} $$

Equation (4) holds for $[\tau - \psi(s)] \geq 0$. It is obvious that $\varphi(0) = 0$. It follows from Eq. (4) that:

1. For s = const, we get the equation of motion of a point, with a fixed value of s, along the y axis;

---

[*]The author expresses his gratitude to Yu. I. Likhachev for his aid in the calculations.

10

2. For y = const, Eq. (4) gives the equation of the outline of the flaw at the instant of time τ;

3. The condition for the fault occurring all at one time for any τ = const is given by

$$y = \varphi\left[\tau - \psi(s)\right] = \text{const.} \tag{5}$$

Differentiating Eq. (5) with respect to s, we find

$$\frac{d\varphi}{d\psi}\frac{d\psi}{ds} = 0. \tag{6}$$

But

$$\frac{d\varphi}{d\psi} = \frac{dy}{d\tau} = v_y \neq 0,$$

since for $v_y = 0$, the flaw would not open up at all.

Thus, $d\psi / ds = 0$, but

$$\frac{d\psi}{ds} = \frac{1}{\dfrac{ds}{d\tau}} = \frac{1}{v_x} = 0.$$

Accordingly,

$$v_x = \infty. \tag{7}$$

Thus, even for this more general case, if the failure occurs all at one time, it means that the rate of development of the flaw through the cross section of the sample is infinite.

Degree of Localization

In the study of residual stresses, and recently, in failure studies as well, three degrees of localization are distinguished:

1. Macroscopic localization, or localization of the first kind, in zones commensurate with the volume of the sample or part.

2. Microscopic localization, or localization of the second kind, in zones commensurate with the grain dimensions.

3. Submicroscopic localization or localization of the third kind, in zones commensurate with the interatomic distances in the solid.

One must also distinguish degrees of localization in dealing with temperature effects (see Table 2). Thus, for example, in plastic deformation, particularly with repeated loading and under external friction, a very large local temperature rise can occur, reaching hundreds of degrees, although the mean temperature of the body may change very little. Of course, the ratio of mean to local temperature depends on both the size of the heated zone and the thermal properties of the material (such as thermal conductivity, heat capacity, etc.).

A characteristic feature of microtemperature stresses (of the second kind) is that they are of a volume nature, which follows from the volume nature of free thermal expansion.

An estimate (even an approximate one) of the degree of localization makes possible a more accurate delimitation of where the various laws apply, thus avoiding many misunderstandings and inaccuracies.

Thus, for example, if a fatigue fracture is said to be a brittle fracture, what we have in mind is the macroscopic brittleness, since on the microscopic scale these fractures involve considerable plastic deformation, often amounting to tens of percent.

Initial Defects and Damage Done

As the stressing used in present-day construction has been on the increase, engineers have been departing more and more from the apparently unshakable principle, dating from the beginning of the twentieth century, that normal operation must be kept in the elastic region. Rather, at the present time it is difficult to find heavily stressed parts in some of the new branches of engineering which are operating solely in the elastic region, since, to some extent or other, under operating conditions they are almost always seen to enter the plastic region. Of course, when this occurs, it is almost always local (only in certain zones), and moreover limited in magnitude. Until recently, however, it had been assumed to be self-evident that the next stage of the process — failure under operating conditions — was absolutely intolerable, since even the beginning of failure usually produces a large reduction in strength and reliability. However, even this point of view, although it seems obvious, has recently been undergoing revision principally for the reason that it has been found that it is impossible to avoid having defects (in the broad sense of the word — lack of the proper structure) in any real solids.

In essence, dislocation theory is the study of submicroscopic defects in the structure of crystal lattices. The question then is simply whether or not certain defects are allowable. Hence, in particular, it follows that in making failure calculations, instead of continuous media, one starts with media already containing flaws. Using materials of this sort makes it possible to show by calculation that it is dangerous to go from uni- to biaxial tension with brittle solids, although the earlier theory of strength showed no difference between uni- and biaxial tension, and in the newer theory, the strength would turn out to be even greater under biaxial tension than under uniaxial tension, which is at variance with operating experience with pipes and tanks made of low-plasticity materials. Hence, it follows that it is dangerous to have biaxial tension, for example, with rapid local cooling of the surface.

Hardening and Dehardening

As the load increases (or as the time it acts becomes longer), there are various processes that develop and interact with one another in the deformed solid, such as elastic and plastic deformation, flaw formation, and a variety of structural changes, in particular, reduction in grain size, approach to a predominant orientation, precipitation from solid solutions, growth of precipitates, phase transitions at elevated temperatures, and, in chemically active media, corrosive processes as well.

In spite of the fact that these processes are radically different in nature, as far as the overall* effect on the strength of the deformed solid is concerned, any deformation process may, for convenience, be divided into two stages: the stage where hardening predominates, and the stage where dehardening predominates.

This division shows up most clearly in simple tension of a smooth sample. Here, the hardening (increase in resistance per element of volume) is caused by elastic (Hooke's law) and plastic deformations, as may be seen from the curves showing the true stresses (i.e., calculated for the actual, rather than the initial cross section). Sometimes the hardening may be accompanied by such structural processes [11] as, for example, transition from austenite to martensite in cold deformation of austenitic steels, precipitation of finely dispersed carbides in deformation of tempered and low-relaxation steels, etc. At some stage in the deformation, dehardening processes begin to act in addition to the continuing hardening processes, and these reduce the strength or area of some of the loaded cross sections, and thus weaken the sample as a whole. Such processes include the development of plastic neckdown, flaws, and, in some cases, structural processes (for example, growth of structural phases).

Control of the Kinetics of Deformation and Failure

Consider first mechanical loading. In this case, slowing down of plastic deformation and of failure can follow either the same or opposite paths. A fundamental way of slowing down the formation of a fault is to increase the amount of plastic deformation that it encounters in its path. It is desirable, it is true, for this to apply only to the local plasticity. It is still not clear whether or not it is possible to increase the local plasticity and reduce the overall plasticity at the same time.

---

*That is, not the effect on the properties of the material alone, but on the properties of the whole body, including physical and geometric factors. Thus, weakening of the body may be caused either by growth of structural phases (without changing the cross section), which weakens the material, or by reduction in the cross section as a result of "neckdown" or flaw formation (without changing the properties of the material), which constitutes weakening of the body.

Increasing the overall plasticity can, in some cases, considerably increase the strength (bearing power), while in other cases it has practically no effect. The first cases include statically indeterminate systems, in which plastic deformation often produces a large and favorable redistribution of stresses (change in the force state), as well as nonuniformly stressed elements* (in bending and torsion, where stress concentration is present, or where bending and concentration are acting together, e.g., in tension, where notched samples are cut through, etc.).

If, however, there is no static indeterminacy, increasing the plasticity does not produce an increase in strength (e.g., in the case of a rod in tension from a suspended load. Increasing the plasticity in such cases simply increases the residual deformation before failure, but does not increase the bearing power. However, in actual assemblies where there is generally external or internal static indeterminacy, increasing the plasticity can substantially raise the strength. On going to thermal loading, we encounter mechanical stresses and deformations in conjunction with temperature fields. In addition to stresses and deformations, the temperature fields also (with a nonlinear temperature gradient), produce a change in the mechanical and physical properties. It is often possible to produce a greater increase in strength by simply changing the temperature fields than is possible by purely mechanical effects.

Further Problems in the Field of Thermal Strength

It follows first of all from what has been said that it is necessary to take account of the change with time (during loading) of both the mechanical and thermal loads, and the resistance of the parts, since usually neither of these quantities remains even approximately constant [1]. Oscillograms must be taken against time when there are rapid changes in the loads and resistances. In considering the strength change with time, use is usually made of the concept of harm done (damage), which changes from an initial (undamaged) state to a state where the damage is maximum (at the instant complete failure occurs)[12].

It should be emphasized that stressed and deformed states can be investigated with greater accuracy than the danger inherent in the states can be evaluated. In other words, of the two basic steps in evaluating the strength — (a) finding the stressed and deformed states, as well as the corresponding temperature fields, and (b) evaluating the danger, i.e., making an actual strength calculation — the second step has been subject to much less investigation than the first. Some results of creep calculations are given below, in the paper by B. F. Shorr (see this collection, page 116), thermal fatigue results are given in the paper by N. D. Sobolev and V. I. Egorov (see the present collection, page 62), and thermal strength results are given in the paper by L. A. Shapovalov (see present collection, page 159). Strength calculations can be made more accurate by using statistical methods. As is done in the case of mechanical strength, temperature effects and the resistance to them must be evaluated from their probability of occurrence. What should be shown, for example, is that there is a 50% probability of having temperature gradients up to 300°C, 20% probability up to 400°C, and 70% probability up to 200°C, etc. Since there will inevitably be defects of some kind in any actual part, and there is only a small probability that these defects will occur in the danger zone of a small number of test samples, and there is even less chance in parts where only one or two samples are usually tested, it is desirable to make a special evaluation of the sensitivity to flaws [2].

It is known from practical experience with steam pipes that large transient thermal effects occur under operating conditions, especially at start-up and when conditions are being changed, and that these can result in failure of welded joints, particularly when the wall thicknesses are greater than 20 mm. These joints show various defects, in particular intercrystalline faults, which occur either in the start-up period, or in the course of tens of thousands of hours of operation. This shows the need of evaluating sensitivity to flaws under thermal effects, and of working out design and construction methods for reducing the sensitivity. It is a good idea not to limit the defectoscopy to the initial state, but to make periodic defectoscopic checks after definite periods of operation (kinetic defectoscopy). Analyzing and comparing the breaks after test and service failures can be important as well. The objective methods of analyzing breaks, developed in recent years, may be useful for this purpose [13].

---

*A nonuniformly stressed body may be regarded as internally indeterminate.

In spite of all the measures mentioned above, the difference between operating conditions and laboratory tests is still great. Accordingly, high temperatures [14] and natural micromechanical methods should be used in a number of cases. Natural micromechanical methods are those in which initially stressed microscopic samples are placed inside of an operating boiler, reactor, etc. In such cases, the testing machine is, by the nature of the case, absent, and is replaced by a setup of the type used in the study of corrosion cracking. The physical state of the microscopic sample must be kept as close as possible to that of an actual part (with the exception, of course, of the absolute dimensions). Thus, the microsample may be cut out near the welded joints of actual parts, thus keeping the surface in the way in which it has been treated.

There are also various cases of weakly relaxing thermal stresses, for example, where they are of small extent, and, hence, the relaxational motion of the hot zones is small. Since the thermal expansion is only slightly dependent on the temperature, the thermal motion in a nonuniformly heated body will be almost the same as in one that is uniformly heated. However, there will be less removal (by relaxation) of the temperature stresses, the smaller the extent of the heated, and thus strongly relaxing, zones. It should be emphasized that it is necessary to make an experimental study of the temperature fields, since the calculated fields used in various practical cases are often too idealized.

It may be assumed to have been proved in the field of mechanical strength that even the most improved testing of samples cannot completely replace "natural tests" of parts and assemblies, nor observations of how they behave under operating conditions. This is completely true of thermal strength as well.

Accordingly, in addition to sample testing, methods must be developed for finding the "structural thermal strength," i.e., tests for thermal fatigue, thermal shock, thermal stability, etc., of constituent, particularly welded elements, models, and structural assemblies, together with a study of the effect of coatings, surface state, and notches.

An important role in developing the tests is played by new methods of producing transient temperature fields, and are above all capable of producing rapid heating and cooling. It has recently been proposed to make rapid temperature changes by immersing the sample in a powder that has a stream of gas blowing through it [15]. Depending on the temperature of the powder, this method can produce both rapid heating and rapid cooling. The materials used in the powder are chromium, tungsten, silicon carbide, etc.

The thermal conductivity of the materials has little effect, since, in view of the small dimensions of the powder particles (about 0.07 mm), the temperature gradients are small. The gas gets into the powder through a perforated heater made, for example, out of a sheet of stainless steel with a number of 1-mm holes drilled in it. The gas blowing through the powder (fluidized bed) makes it look like a boiling liquid. It is probable that similar methods can be used for both thermal fatigue and thermal shock tests [16].

Summary

A discussion has been given and a comparison made between some of the laws that are of importance to mechanical and thermal strength, i.e., the laws governing flow, creep, one-time static and fatigue failure, as well as loss of strength under mechanical stresses on the one hand, and under thermal stresses on the other.

1. It is proposed that four types of loading be distinguished: shock load and the three kinds of static loading — short-time, long-time, and repeated.

It is emphasized that true shock loading (not mocked-up statically) is greatly affected by wave processes and inertial forces, and that the characteristics which are not substantially affected by wave processes or inertial resistances (e.g., the shock viscosity), may in principle be replaced by static characteristics (e.g., by tests on notched samples). In laboratory tests and under operating conditions, static processes can occur under a fixed load, and with a fixed displacement and, in the general case, with varying loads and displacements. Thermal stresses usually occur with fixed displacements, and thus relax with increase in deformation, while mechanical stresses may be nonrelaxing.

2. As the magnitude and duration of the loading increase, elastic and plastic deformations are superimposed on one another, and failure occurs. Each of these three regions exhibits both gradual and discontinuous losses of strength. Examples of gradual loss of strength are provided by the elasticity and flow limits, as

determined by the "tolerance" and the hardness under indentation, as well as by the limits on the number of flaws. These usually do not lower the bearing strength. Examples of sudden strength loss are provided by the critical loadings for loss of strength in the elastic and plastic regions, as well as by the stresses when a yield peak or jags are present, and the ultimate stresses in short-time, long-time, and repeated loadings.

3. The failure process results for a longer time the more nonuniform the loading (e.g., if the sharpness of notches is increased, the relative extent of the failure range increases, reaching 90% or more of the total life-time) and the more nonuniform the structure and properties of the material.

Therefore, under operating conditions (where there are notches, and the loading and structure are nonuniform), failure and the postcritical state of parts usually arise considerably earlier, and hence actual testing is of greater importance than laboratory tests of samples having considerably less nonuniformity.

4. The frequent practice of taking the total time of the process (e.g., until formation of flaws or until complete failure occurs) is not good enough to distinguish between gradual and sudden losses of strength, or to find the kinetic laws that apply. These laws are observed when curves are constructed showing the acceleration j_ε of the deformation, and j_{tf} of failure, in particular, since the precritical state (j_ε and j_{tf} less than zero) differs from the postcritical state, or fourth kinetic period (j_ε and j_{tf} greater than zero).

In flow, creep, and failure, it is proposed to distinguish four periods in the general case: the incubational or initially accelerated period ($j > 0$, $v > 0$), the slowing down period ($j < 0$, $v > 0$), the steady-state or uniform development period ($j = 0$, $v = 0$ or const), and the final accelerated period (sometimes avalanche-like with $j > 0$, $v > 0$). It is shown that there is an incubation period even in creep, which shows the common error of dividing the creep process into three periods, neglecting the incubation period.

5. The need for a kinetic approach is due to the essentially transient character of the majority of practical cases of loading. The transient character may be produced by a change in either the external factors (e.g., temperature gradients in start-ups, shutdowns, and changes of operating level), or the internal factors (e.g., structural changes or physicochemical processes occurring under constant external load).

6. The need is emphasized for taking account of localization, and making a distinction between processes occurring under thermal load involving deformation and failure of the first (macroscopic), second (microscopic), and third (submicroscopic) types. Evaluation of the phenomenon is essentially dependent on the degree of localization; for example, submicroscopic failure can show up averaged as a macroscopic plastic deformation.

7. In making failure calculations, the ideas of a continuous medium taken from the theory of elasticity and plasticity must be replaced by a model of the body with its original defects (cracks) existing even before loading. Making this substitution leads to a more correct calculation of many phenomena (e.g., the danger from biaxial tension in brittle materials, conditions under which flaws are nondeveloping).

8. Under constant load, the acceleration of the process may be used as a criterion: in hardening, the process is slowed down (i.e., $j < 0$) because of the increase in resistance with time resulting from cold working or structural transformations, and in dehardening ($j > 0$), the process is accelerated either as a result of weakening of the material, as a result of the true cross sections being reduced by plastic deformation or formation of flaws.

9. Control of the deformation and failure kinetics (mainly slowing down these processes) may be accomplished by a change in either the external factors (loads, temperature fields, forms and conditions of hardening), or the internal factors (structure, composition, coatings, etc.). Increasing the elastic energy supply of the loading system leads to slower relaxation of the load with increase in deformation, and thus to a more sudden, and sometimes even earlier occurrence of plastic deformation and failure and of the critical states.

10. The effect which plasticity has on the strength of parts shows up as a change in the force (redistribution of the forces in statically indeterminate systems) and in the stressed states (in equalizing the initial nonuniformity of the stresses and breaking, e.g., from having an initial cut-through, with stress concentrations). In statically indeterminate systems with no large initial nonuniformity, plasticity has no essential effect on the strength (if microscopic processes are neglected).

11. In order to get better agreement between laboratory tests and transient operating conditions, laboratory methods must be developed for finding the mechanical properties under transient conditions, in particular under repetitive temperature changes (thermal fatigue), under sudden thermal effects (thermal shock), as well as when there are various combinations of mechanical and thermal effects, with notches, welded joints, and other factors present which produce a nonuniform and transient loading. The possibility is noted of using a new method of rapid heating and cooling by immersing the sample in a fluidized bed, which has a stream of hot or cold gas blowing through it.

12. Since it is usually impossible to get rid of plastic deformation and early flaws in present-day assemblies under high mechanical and thermal loading, the fundamental problem is to slow down (hinder) these processes, and thus keep them from getting to the critical, or even more to the postcritical state.

LITERATURE CITED

1. Fridman, Ya. B., Zilova, T. K., Drozdovskii, B.A., and Petrukhina, N. I. "Calculation of mechanical characteristics including deformation and failure kinetics," Zavod. Lab. (11): 1267 (1960); Vagapov, R. D., and Fridman, Ya. B., "Comparison of strength under fixed load and fixed displacement," Zavod. Lab. (2):183 (1961).
2. Drozdovskii, B. A., and Fridman, Ya. B., Effect of Flaws on the Mechanical Properties of Construction Steels, Moscow, Metallurgizdat, 1960; Some Problems in the Strength of Solids, Moscow, Izd. AN SSSR, 1959, p. 280; Zavodsk. Lab. (3):320 (1959).
3. Fridman, Ya. B., Mechanical Properties of Metals, 2nd ed., Moscow, Oborongiz, 1952, p. 383.
4. Fridman, Ya. B., in: Testing Machine Parts for Strength, Moscow, Mashgiz, 1960, p. 3.
5. Oding, I. A., Ivanova, V. S., Burdukskii, V. V., and Geminov, V. N., Theory of Creep and Long-Time Strength, Moscow, Metallurgizdat, 1959, p. 109.
6. Sally, A., Creep of Metals and Heat-Resistant Alloys, Moscow, Oborongiz, 1953, p. 118.
7. Arskii, V. N., Fiz. Metal. i Metalloved 6(3):496 (1958).
8. Tapsell, Creep of Metals, ONTI, 1934.
9. Drozdovskii, B. A., and Fridman, Ya. B., Dokl. Akad. Nauk SSSR 95(4):793 (1954).
10. Likhachev, V. A.,et al., Nauchn. Byul. Lening. Politekhn. Inst. Razd. Fiz.-Matem. Nauk, No. 12 (1958).
11. Kishkin, S. T., Izv. Akad. Nauk SSSR, Otd. Tekhn. Nauk (12):1799 (1946); (1):87 (1948).
12. Sobolev, N. D., Fiz. Metal. i Metalloved. 9(5):758 (1960).
13. Fridman, Ya. B., Gordeeva, T. A., and Zaitsev, A. M., Structure and Analysis of Breaks in Metals, Moscow, Mashgiz, 1960.
14. Konoplenko, V. P., and Vinogradov, D. K.,Zavodsk. Lab. (1):106 (1959).
15. Glenny, E., Northwood, J., Shaw, S., and Taylor, T., J. Inst. Metals (London) 87:274 (1959).
16. Fridman, Ya. B., Sobolev, N. D., Borisov, S. V., Egorov, V. I., Konoplenko, V. P., Morozov, E. M., Shapovalov, L. A., and Shorr, B. F., "Some thermal strength problems in reactor construction." At. Energ. 10(6):606 (1961).

THERMAL STRESSES AND THEIR CALCULATION

E. M. Morozov and Ya. B. Fridman

Thermal (or temperature) stresses are a variety of internal (inherent) stress. This name, as we know, is given to stresses that are in equilibrium in a given body (or part of it, or in a system of several bodies), with no "external" loads, such as, for example, mounting loads or surface stresses, which sometimes occur in hardening treatments given to part surfaces, or the residual stresses formed after a previous nonuniform plastic deformation has been removed. Usually, internal stresses are divided according to the degree to which they are localized [1]:

Stresses of the Zero Kind. These forces arise in a system of mutually connected bodies (including bonds applied to a body) as a result of their combined effect on one another either from nonuniform thermal action on the bodies making up the system (e.g., in shrink fitting of a sleeve on a shaft), or from having different coefficients of linear expansion (e.g., in changing the temperature of bimetallic pairs).

Stresses of the First Kind (Macroscopic). These forces come into equilibrium in regions of the order of the dimensions of the body. They are caused by nonuniformity in the force and temperature fields, or in the properties of the body.

Stresses of the Second Kind (Microscopic). These forces come into equilibrium in regions of the order of the grain dimensions of the material, for example in heating and cooling of two micrograins joined together, with different coefficients of expansion α_1 and α_2, and elasticity moduli E_1 and E_2. Here [2], we have

$$\sigma = (\alpha_1 - \alpha_2) \Delta T \frac{E_1 E_2}{E_1 + E_2}.$$

Distortions* of the Third Kind (Submicroscopic). These distortions come into equilibrium in regions of the order of the dimensions of the interatomic distances. They are still almost unstudied and are not discussed in this review.

A distinction is made between two types of internal (inherent) stresses: temporary, or reversible, i.e., those which disappear after the cause is removed, and residual or irreversible, i.e., those that remain after the cause is removed (e.g., the residual stresses formed when the temperature is equalized after the stresses have passed beyond the flow limit).

Thermal stresses differ from other kinds of internal forces only in what causes them. Just as with other types of internal stresses, thermal stresses are divided into:

a. Zero, first, and second order

b. Temporary and residual

c. Thermoelastic and thermoplastic

---

*They cannot be called stresses, since when talking about very small zones it is impossible to use the concept of stress as it is used in the mechanics of continuous media.

In addition to purely physical (reversible) thermal expansion, a change in temperature produces effects in many cases which lead to thermal macro- and microstresses, for example, as a result of phase transformations, accompanied by volume changes (both in the solid state — in cementation by a solid carburizer, and when the state of aggregation changes — in crystallization, etc.). Accordingly, nonuniform temperatures produce irregular volume and linear phase changes, and hence produce internal stresses. Other causes of stress are difference in relaxation rate, or difference in the rate of physicochemical processes occurring in different parts of the body when they are unequally heated, for example in drying wood fiber and in corrosion. Sometimes phase stresses are called stresses of the fourth kind or nonloading stresses. There is, however, no necessity for these terms, since what we are dealing with are internal phase stresses of the first or second kind.

The reason thermal stresses arise is that thermal expansion is restricted in the body or in some of its parts. Free homogeneous solids with a uniform or linear change in temperature over the solid undergo free thermal expansion (thermal deformation), which does not produce any macroscopic stresses. Thermal deformations are seldom free under operating conditions.

Separate but interconnected parts of a body may be either differently heated as a result of a temperature gradient, or they may have different coefficients of expansion as a result of anisotropy or inhomogeneity in properties (for example in heterogeneous solids). In both cases there is a difference in the temperature deformation αT produced at different points or in different directions in the solid, and thus some of the parts can interfere with expansion of the others as a result of the restraint supplied at the boundary. For example, a less strongly heated part of the solid will restrict the deformation of a part that is more strongly heated. Just as with other forms of internal stresses, when thermal stresses interact with the stress field from the external load, they can be dangerous in some cases, while in others they can be useful in increasing the strength and operability of parts and equipment.

Starting with a continuous material, the theory of elasticity assumes that the geometric deformation of an isolated element of volume is equal to the sum of the elastic (found from the generalized Hooke's law) and the thermal deformations

$$\frac{\partial u}{\partial x} = \frac{1}{E}\left[\sigma_x - \mu\left(\sigma_y + \sigma_z\right)\right] + \alpha T;$$

$$\frac{\partial u}{\partial y} = \frac{1}{E}\left[\sigma_y - \mu\left(\sigma_x + \sigma_z\right)\right] + \alpha T;$$

$$\frac{\partial w}{\partial z} = \frac{1}{E}\left[\sigma_z - \mu\left(\sigma_x + \sigma_y\right)\right] + \alpha T.$$

The following notation is used here and in the rest of the book:

$\alpha, 1/°C$ . . . . coefficient of linear temperature expansion;

$E, kg/cm^2$ . . . . modulus of elasticity and tension;

$T = T_f - T_i$ . . . . temperature drop at a given point in the body from T_i, in the original unstressed body, to T_f, the temperature after heating;

μ . . . . Poisson's ratio;

$\sigma_x, \sigma_y, \sigma_z$ . . . . normal stresses on areas, the normals to which coincide with the directions of the coordinate axes x, y, z. If, at the same time, the tangential stresses on these areas are equal to zero, the stresses $\sigma_x, \sigma_y, \sigma_z$ become the principal stresses, and will have the extremum values for the given point in the body;

$\tau_{xy} = \tau_{yx}, \tau_{zy} = \tau_{yz}, \tau_{xz} = \tau_{zx}.$ tangential stresses on mutually perpendicular areas, the first subscript gives the direction of the normal to the area, and the second gives the direction of the vector τ;

$\sigma_\Theta, \sigma_r, \sigma_z, \tau_{r\Theta}$ . . . . normal and tangential stresses in cylindrical coordinates;

e_x, e_y, e_z . . . . relative linear deformations along the coordinate axes x, y, z;

$\gamma_{xy} = \gamma_{yx}, \gamma_{xz} = \gamma_{zx}, \gamma_{yz} = \gamma_{zy}.$ relative angular deformations (distortions of the right angle between the directions indicated);

u, v, w	absolute displacement of points in the body along the axes x, y, z;
r	variable radius in polar coordinates;
a, b	inner and outer radii of a tube or annular disc;
t	time;
δ	plasticity in tension;
σ_T, σ_b, kg/cm^2	flow limit and ultimate strength of material in the body, respectively;
λ, cal/cm-sec-deg	thermal conductivity;
q, cal/cm^3-sec	specific power (intensity) of heat generation;
Q, cal/sec	total heat power generated or supplied;
c, cal/g-deg	specific heat capacity;
γ, g/cm^3	density.

The temperature distribution and corresponding stresses and displacements are found from successively solving the equations of heat conduction and those of the theory of elasticity (or plasticity).

The temperature drop may be produced either by a heat flux from internal heat generation, or by a temperature difference on the boundary surfaces of the body. In the case where the thermal stresses reach the flow limit, the body, or parts of it, pass to the plastic state, and thermoplastic stresses arise which accordingly become residual stresses after the load is removed. The residual stresses are defined as the difference between the acting elastoplastic stresses, and the fictitious elastic stresses which would occur (at the same thermal loading), if there were no plastic deformation.

Thus, three basic types of thermal stresses may be distinguished:

1. Thermal Elastic (or Thermoelastic) Stresses. These stresses occur in an elastic solid (which is undergoing neither plastic deformation nor phase transformations). They are temporary, i.e., they disappear after the load is removed. In the vast majority of cases, these are the only stresses that are given by calculation, and the greater part of the results given below reflect these stresses alone.

2. Thermoplastic Stresses. These stresses occur when a heated body passes to the plastic state. In some (so far not many) cases, these stresses may also be found by calculation.

3. Thermal Phase Stresses. These stresses result from nonuniform (in the degree or velocity of phase transformation) transition of an irregularly heated solid to another phase state. So far, these stresses have only seldom been found by calculation, although they are of great practical importance when materials undergo phase transformations with large volume changes, for example in iron, steel, uranium, plutonium, and their alloys.

Stresses of all these types can occur in solids which are homogeneous or inhomogeneous, anisotropic, etc., in either their physical or mechanical properties.

In the elastic region, the stresses at any instant of time are determined by the existing temperature field (as long as the temperatures are not changing too rapidly with time), while in the plastic region the stresses also depend on the "prehistory" of the solid. With very rapid temperature change (thermal shock), both the magnitude and rate of change of the forces acting are of fundamental importance to the strength of the solid.

Actual solids are scarcely ever ideally elastic, since they are to some degree viscoelastic or viscoplastic, the more so the higher the temperature. Accordingly, both the thermoelastic and the thermoplastic stresses relax (decrease) like other forces in course of time, as a result of increase in creep deformation, at the expense of reducing the elastic part of the total deformation, and this is more likely, the higher the temperature [3].

In a system of solids, or in a restrained solid, thermal stresses of the zero and second kinds can occur for any temperature field.

In a free (unclamped) solid, thermal stresses of the second kind can also occur for any temperature field, while thermal stresses of the first kind occur only in a temperature gradient that is not linear.

If thermal stresses of the first kind are absent, the free thermal deformation, i.e., the deformation which is not acted against by resisting forces, will be equal to

$$\varepsilon_x = \varepsilon_y = \varepsilon_z = \alpha T,$$

$$\gamma_{xy} = \gamma_{xz} = \gamma_{yz} = 0.$$

If these expressions are substituted into the equations for the compatibility of deformations (familiar from the theory of elasticity), we obtain

$$\frac{\partial^2 T}{\partial x^2} + \frac{\partial^2 T}{\partial y^2} = 0; \quad \frac{\partial^2 T}{\partial y^2} + \frac{\partial^2 T}{\partial z^2} = 0; \quad \frac{\partial^2 T}{\partial z^2} + \frac{\partial^2 T}{\partial x^2} = 0;$$

$$\frac{\partial^2 T}{\partial x \, \partial y} = 0; \quad \frac{\partial^2 T}{\partial y \, \partial z} = 0; \quad \frac{\partial^2 T}{\partial x \, \partial z} = 0.$$

The solution of this system of equations gives the temperature distribution law for which thermal stresses of the first kind are absent:

$$T = a_0 + a_1 x + a_2 y + a_3 z.$$

This shows that thermal macrostresses do not occur in a free body with a linear temperature gradient.

The following formula may be used to find the order of magnitude of the thermoelastic macrostresses (of the first kind)

$$\sigma = \frac{\alpha E T}{1 - k\mu} k_f k_\tau,$$

where k_f is the form factor, taking on different values depending on the geometry of the solid, k_T is the temperature factor, always less than or equal to unity, depending on the nature of the thermal load applied (rate of change, etc.), and k is the clamping factor.

Depending on the shape of the solid and the way it is clamped, the clamping factor has the following values:

Linear solid (rod), clamped at the ends 0
Flat solid (plate), clamped at the edge 1
Three-dimensional solid (in bulk), clamped at the surface 2

The symbol T denotes the temperature difference at a given point in the solid:

$$T = T_f - T_i.$$

It is usually assumed in subsequent problems that the solid is heated, i.e., $T_f > T_i$, and hence T > 0. If the solid is being cooled, all the formulas remain the same, but the sign of T is reversed. Finding the sign of the thermal stresses produced is illustrated for a sphere in Table 1. With rapid heating, a temperature gradient occurs as a result of the transient nature of the process, while with very slow heating there is no temperature gradient, and no stresses occur. With rapid cooling, the signs are reversed.

To find what effect the nature of the material has on the magnitude of the thermal stresses (other conditions being equal), we give values of the stress αE (kg/cm^2), produced by one degree of heating of an end-clamped rod made of the material given in the table.

The values for several materials are as follows:

Steel 45 . . . . 24.7
Heat-resistant steel ÉI437 . . . . 25.2
Stainless steel ÉI69 33
Aluminum 16.7
Magnesium 11.7
Graphite 0.14-0.21
Thorium 7.85
Uranium 46.2

TABLE 1. Signs of the Stresses in the Outer and Inner Zones
of a Rapidly Heated Solid

Elastic or elastoplastic stresses in rapid heating	
Residual stresses after equalizing the temperature (in case the elastoplastic state occurred during heating)	

TABLE 2. Coefficient of Linear Expansion α of Various Elements

Element	Crystal structure	$\alpha \cdot 10^{-6}$, 1/°C	Element	$\alpha \cdot 10^{-6}$, 1/°C
Uranium	α-U	36.7 (100) −9.3 (010) 34.2 (001)	Thorium	11.15
	β-U	4.6 ($\parallel$) 23 ($\perp$)	Beryllium	12.4
			Molybdenum	5.5
			Aluminum	23.9
	γ-U	18	Magnesium	26
Plutonium	α-Pu δ-Pu	55 −30	Graphite	2-3
Zirconium	α-Zr	10.8 ($\parallel$) 5.6 ($\perp$)		

If thermal stresses are to be reduced, preference must be given to the materials with the smaller values of αE.

The physical and mechanical properties of several materials [3,4] are given in Tables 2 and 3.

In calculating the stresses in the elastic range, the change in α and E with increase in temperature is usually neglected, since transition to the plastic range generally occurs before α and E change appreciably.

In case it is desirable to include the change in the coefficient of thermal expansion with temperature (α usually increases with T); the general formulas in which the stresses are expressed as integrals of T must have α under the integral sign, so that the integrals are taken of the product αT instead of T alone.

If the boundary conditions in any given problem do not fit the type of clamping on the edges of a body in the problems given below, the form that we are interested in may be obtained from the one given by superimposing the solution known from the theory of elasticity which corresponds with the boundary conditions needed.

In applying the solutions of problems for infinitely large or infinitely long solids to solids of finite dimensions, we must keep in mind Saint-Venant's principle, since the stresses found for an infinite solid will be distorted at the ends of the actual solid. From Saint-Venant's principle, for tubes, the distortion extends a distance of two or three wall thicknesses from the end of the tube, while for strips and beams it extends a distance of the order of the largest cross-sectional dimension of the rod, and, for an infinite or semi-infinite solid, the influence of the boundaries of an actual finite solid has no effect beyond five times the largest diameter of the perturbed region.

TABLE 3. Mechanical Properties of Various Materials at Different Temperatures

Properties of the material	Temp., °C	Uranium	Thorium	Zirconium	Beryllium	Molybdenum	Aluminum	Magnesium	Graphite	Stainless steel
$E \cdot 10^{-6}$, kg/cm²	20	1.05-1.75 (stat.) 2.1 (dyn.)	0.7	1	2.9	3.15	0.7	0.45	0.07	2.1
σ_T, kg/mm²	20	15.6	19	12.5	23	40-60	10-15	10-23	–	31.5
	300	12.7	8.4	7	10.5	–	1.05	3.5	0.1	21
	500	4.2	–	–	–	–	–	–	–	–
σ, kg/mm²	20	63.3	26.6	46	29	70-175	15	17-32	–	63
	300	22.5	15.4	27	22	42-56	2	6.6	–	45.5
	500	7	–	20	20	–	–	–	–	–
δ, %	20	13.5	40	16-31	2	5-20	35	5-15	–	55
	300	43	38	60	17	–	20	23	–	40
	500	57	–	–	–	–	–	–	–	–
μ	20	0.23	0.26	0.33	0.03	0.31	0.3	0.35	0.2-0.33	0.38
λ, cal/cm · sec · deg	20	0.06	0.09	0.05	0.38	0.53	0.53	0.35	0.15	–

Notches concentrate both thermal stresses and deformations as well as the stresses from external loads.

This paper gives a short compilation of the results found in the literature for thermal stress calculations in various cases. Having engineering applications in mind, we have limited ourselves to giving for each case only those results which are most needed for practical purposes.

Unfortunately, knowing only the magnitudes and distributions of the temperature stresses is not enough to ensure strength and reliability of construction. We also need data to evaluate the danger from thermal stresses and to determine what measures should be taken to reduce the danger.

These problems are still far from being completely solved, and they must be considered in conjunction with the mechanical strength of the materials and the designs (creep resistance, long-time static failure, fatigue, loss of strength, etc.). Thus, for example, while a single rise in the temperature stresses is practically without danger in a body that is capable of plastic deformations, which reduce the stress, flaws may be formed in a brittle body, or complete failure may occur.

The data given below may turn out to be useful to people working in design and calculation fields, in production and use of machines, and of buildings and equipment in which temperature stresses occur.

Three-Dimensional Bodies, Spheres, and Half-Spaces

1. Shape of Body [3]

Free body (three-dimensional) of any shape (Fig. 1).

Temperature Effect. Nonuniform heating. No internal heat source. Temperature change law:

$$T = ax + by + cz + d,$$

where a, b, c, and d are constants or functions of the time.

Stresses. No thermal stresses occur in this case. With a body unclamped at the boundaries, a linear temperature gradient does not produce any concentration of the deformations, since the derivative of the temperature with respect to the coordinates is constant.

Deformations. Free, purely temperature deformations:

$$\varepsilon_x = \varepsilon_y = \varepsilon_z = \alpha T,$$

$$\gamma_{xy} = \gamma_{yz} = \gamma_{xz} = 0.$$

2. Shape of Body [5]

Long free body of cylindrical shape, and arbitrary cross section (Fig. 2). The shape of the cross section does not vary along the length. The body is clamped at the ends (no motion along the Z axis).

Temperature Effect. Steady-state temperature distribution T(x,y), satisfying the Laplace equation:

$$\frac{\partial^2 T}{\partial x^2} + \frac{\partial^2 T}{\partial y^2} = 0.$$

The temperature does not change along the length. There are no heat sources or sinks in the region bounded by the outside surface of the body.

Stresses

$$\sigma_z = -\alpha E T.$$

No stresses occur along the other axes ($\sigma_x = \sigma_y = 0$).

Deformations

$$\varepsilon_x = \varepsilon_y = (1 + \mu)\alpha T,$$
$$\varepsilon_z = 0.$$

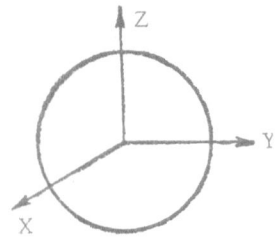

Fig. 1. Free body of arbitrary shape.

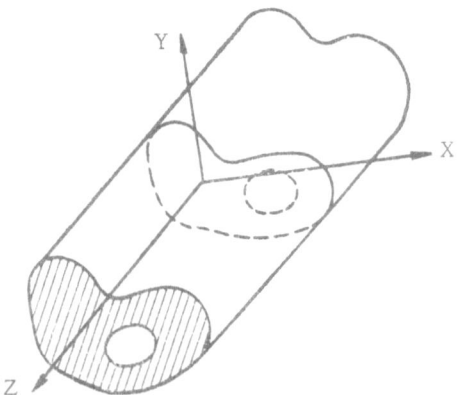

Fig. 2. Cylindrical body, clamped at the ends.

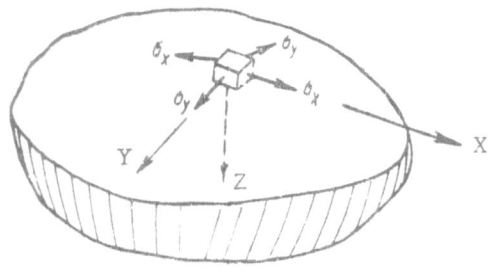

Fig. 3. Stresses in the free surface of a body.

3. Shape of Body [5]

Homogeneous (three-dimensional) solid of any shape (see Fig. 1).

No displacement permitted at any part of the boundary (including holes if there are any).

Temperature Effect. Uniform heating to T degrees.

No heat sources in the volume bounded by the external surface of the body.

Stresses

$$\sigma_x = \sigma_y = \sigma_z = -\frac{aET}{1-2\mu}.$$

Deformations. All deformations (linear and angular) are equal to zero.

4. Shape of Body [6]

Free body of arbitrary shape with any sorts of holes or voids.

Temperature Effect. Nonuniform elevation of temperature with the original temperature uniform:

$$T = T(x, y, z).$$

Volume Deformation. The absolute increase in the volume V of the body is found as the sum of the volume changes in the separate elements of the body, due only to heating, neglecting elastic deformations, since the increase in volume of the body from temperature stresses of the first kind is equal to zero (from the self-balancing property of the thermal stresses in a free body).

$$\Delta V = \int_V 3aT dV.$$

5. Shape of Body [6]

Free body (three-dimensional) of any shape.

Temperature Effect. Part or all of the surface of the body is practically instantaneously cooled from temperature T_2 to T_1.

Stresses. At the initial instant of time (no temperature change yet in the sub-surface layers), there is a biaxial surface tension with stresses equal to

$$\sigma_x = \sigma_y = \frac{aE}{1-\mu}(T_2 - T_1),$$

at the part of the surface where the temperature change occurred (Fig. 3). With the passage of time, the temperature of the body equalizes out, and the stresses decrease.

6. Shape of Body [7]

Isotropic and homogeneous semi-infinite body (half-space).

Temperature Effect. Area in the form of a circle of radius a on the surface (Z = 0) of a semi-infinite body (Fig. 4) heated to the constant temperature T. All other points in the body are at temperature zero.

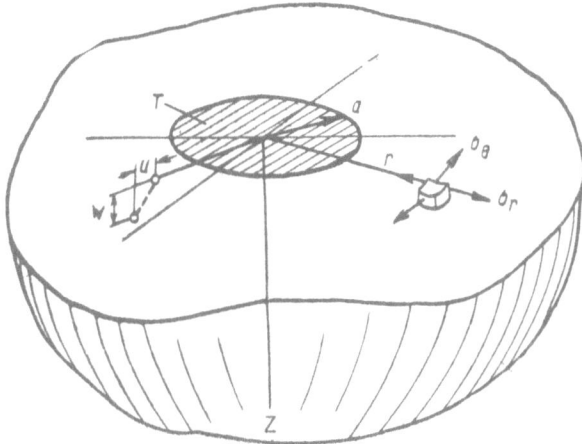

Fig. 4. Stresses and displacement of points on the surface of a body with a heated circular region.

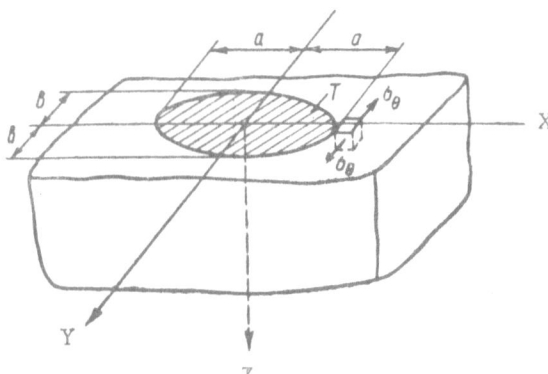

Fig. 5. Heated elliptical region on the surface of a body.

Stresses. The stressed state is symmetric with respect to the Z axis. The principal stresses acting at the Z axis are equal to

$$\sigma_r = \sigma_\Theta = -\alpha ET\left(1 - \frac{z}{\sqrt{a^2 + z^2}}\right).$$

Further from the Z axis, the stresses decrease. At the surface of the body (Z = 0), the stresses will be as follows:

In the heated region (r < a)

$$\sigma_r = \sigma_\Theta = -\frac{\alpha ET}{2}.$$

At the boundary of the heated region (r = a)

$$\sigma_r = -\frac{\alpha ET}{2}, \quad \sigma_\Theta = 0.$$

In the heated region (r > 0)

$$\sigma_r = -\frac{\alpha ET}{2}\frac{a^2}{r^2}, \quad \sigma_\Theta = \frac{\alpha ET}{2}\frac{a^3}{r^3}.$$

The maximum tangential stresses in the unheated region will be on the planes at an angle of 45° to the radius:

$$\tau = \frac{\alpha ET}{2}\frac{a^2}{r^2}.$$

Displacements. The deformed state is symmetric with respect to the Z axis.

The displacement in the Z direction on the Z axis itself is equal to

$$w = \alpha(1 + \mu)\, T\,(\sqrt{a^2 + z^2} - z).$$

The displacement on the surface (Z = 0) at the boundary of the heated region (r = a) is equal to

$$w = \frac{2a(1 + \mu)\, Ta}{\pi}.$$

The radial displacement on the surface of the body will be:

Inside the heated region (r < a)

$$u = \frac{\alpha(1 + \mu)T}{2}\, r.$$

On the boundary of the heated region (r = a)

$$u = \frac{\alpha(1 + \mu)\, Ta}{2}.$$

In the unheated region (r > a)

$$u = \frac{\alpha(1 + \mu)\, T}{2}\frac{a^2}{r}.$$

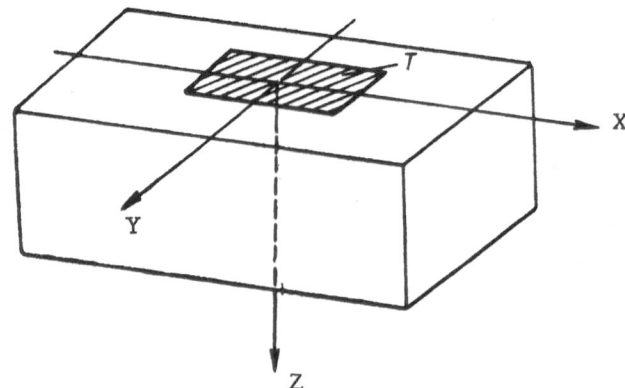

Fig. 6. Heated rectangular region on the surface of a body.

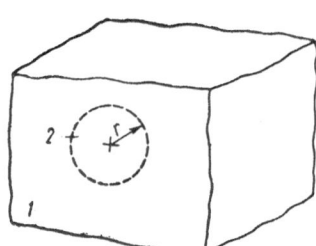

Fig. 7. Spherical inclusion
in a body.

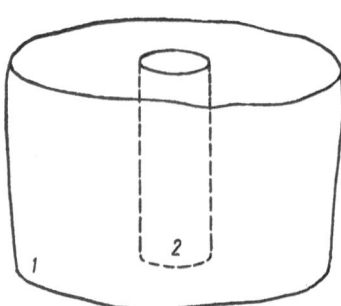

Fig. 8. Cylindrical inclusion
in a body.

7. Shape of Body [6]

Semi-infinite body (half-space).

Temperature Effect. On the surface of the half-space there is a uniformly heated region in the form of an ellipse (Fig. 5). The temperature drop between the heated and unheated regions is equal to T.

Stresses. The maximum normal stress will be at the ends of the long axis of the elliptical region:

$$\sigma_\Theta = \frac{\alpha E T}{1 + \frac{b}{a}} .$$

For a long ellipse ($b/a \to 0$), the maximum stress $\sigma_\Theta \to \alpha E T$.

8. Shape of Body [6]

Semi-infinite body (half-space).

Temperature Effect. The surface of the half space has a region in the form of a rectangle heated to temperature T (Fig. 6).

Stresses. Large tangential stresses occur at the corners, theoretically equal to infinity as the angle becomes more acute. Accordingly, local plastic deformations or flaws are to be expected at the corners. The normal stresses σ_x and σ_y do not reach the value $\alpha E T/2$.

9. Shape of Body [3]

Infinite medium 1 with a spherical inclusion of another material 2 (Fig. 7).

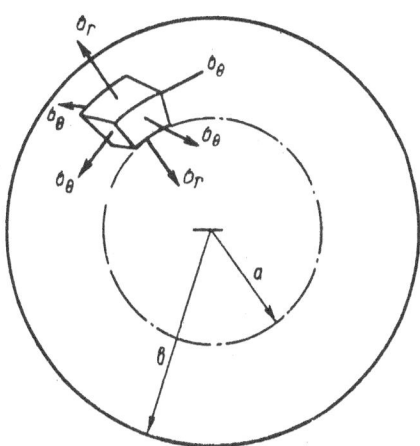

Fig. 9. Stresses in a hollow sphere.

Temperature Effect. Uniform heating of medium and inclusion to the same or different temperatures in such a way that αT is different for the medium and for the inclusion.

Stresses. A radial pressure occurs on the surface of the inclusion and the medium, thus

$$\sigma_r = -\frac{3E_1 E_2 \left[\alpha_1 (T_1 - T_2) + T_1 (\alpha_1 - \alpha_2)\right]}{3E_2 (1 + \mu_1) + 2E_1 (1 - 2\mu_2)}.$$

(the subscript 1 is for the medium, and 2 is for the inclusion).

The maximum peripheral tensile stresses occur in the medium at the junction surface:

$$\sigma_\theta\big|_{\max} = \frac{\sigma_r}{2}.$$

10. Shape of Body [3]

Infinite medium 1 with a cylindrical inclusion of another material 2 (Fig. 8).

Temperature Effect. Uniform heating of medium and inclusion as in the preceding case.

Stresses. At the boundary surface between the medium (subscript 1) and the inclusion (subscript 2), the pressure is equal to

$$\sigma_r = -\frac{9E_1 E_2 \left[\alpha_1 (T_1 - T_2) + T_1 (\alpha_1 - \alpha_2)\right]}{9E_2 (1 + \mu_1) + 2E_1 (1 + \mu_2)(1 - 2\mu_2)}.$$

At the boundary, the peripheral tensile stress in the medium is

$$\sigma_\theta\big|_{\max} = \sigma_r.$$

11. Shape of Body [5]

Hollow sphere. Inner radius a, outer radius b (Fig. 9).

Temperature Effect. The temperature distribution is spherically symmetric about the center of the sphere:

$$T = T(r).$$

Stresses

$$\sigma_\theta = \frac{\alpha E}{1 - \mu}\left[\frac{2r^3 + a^3}{b^3 - a^3}\frac{1}{r^3}\int_a^b Tr^2 dr + \frac{1}{r^3}\int_a^r Tr^2 dr - T\right];$$

$$\sigma_r = \frac{2\alpha E}{1 - \mu}\left[\frac{r^3 - a^3}{b^3 - a^3}\frac{1}{r^3}\int_a^b Tr^2 dr - \frac{1}{r^3}\int_a^r Tr^2 dr\right].$$

The radial stresses are equal to zero on the inner and outer surfaces. The stresses on a solid sphere may be found by setting a = 0.

12. Shape of Body [3]

Hollow sphere. Here, and in what follows, c = b/a (see Fig. 9).

Temperature Effect. The steady heat flux through the wall of the sphere is symmetric about the center of the sphere:

$$T = \frac{T_a - T_b}{c - 1}\left(\frac{b}{r} - 1\right).$$

$$\sigma_\Theta = \frac{\alpha E\,(T_a - T_b)\,c}{(1-\mu)\,(c^3-1)}\left[c + 1 - (cb + b + a)\frac{1}{2r} - \frac{ab^2}{2r^3}\right];$$

$$\sigma_r = \frac{\alpha E\,(T_a - T_b)\,c}{(1-\mu)\,(c^3-1)}\left[c + 1 - (cb + b + a)\frac{1}{r} + \frac{ab^2}{r^3}\right].$$

The peripheral stresses σ_Θ become a maximum at the outside surface of the sphere:

$$\sigma_\Theta\,\text{max} = \frac{1}{2}\,\frac{c^2-1}{c^3-1}\,\frac{\alpha E\,(T_a - T_b)}{1-\mu}.$$

The radial stresses on the inner and outer surfaces are equal to zero, while the maximum stress is reached at a value of r given by the equation

$$r = \frac{\sqrt{3}\,ab}{\sqrt{b^2 + ab + a^2}}$$

If the wall thickness of the sphere is small [so small that we may neglect the square of the quantity m, where m = c − 1 = (b − a)/a], we find for r = a

$$\sigma_\Theta = -\frac{\alpha E\,(T_a - T_b)}{2\,(1-\mu)}\left(1 + \frac{2}{3}\,m\right);$$

for r = b

$$\sigma_\Theta = \frac{\alpha E\,(T_a - T_b)}{2\,(1-\mu)}\left(1 - \frac{2}{3}\,m\right),$$

while the radial stress is equal to zero.

13. Shape of Body [3]

Solid free sphere of diameter 2b.

Temperature Effect. Heat is generated in the material of the sphere at a constant rate q over the whole volume of the sphere:

$$T = \frac{q}{6\lambda}\,(b^2 - r^2).$$

The temperature is assumed equal to zero on the outside surface.

Stresses

$$\sigma_\Theta = \frac{\alpha q E}{15\lambda\,(1-\mu)}\,(2r^2 - b^2);$$

$$\sigma_r = \frac{\alpha q E}{15\lambda\,(1-\mu)}\,(r^2 - b^2).$$

The peripheral stresses become maximum on the outside surface of the sphere:

$$\sigma_\Theta\,\text{max} = \frac{\alpha q E b^2}{15\lambda\,(1-\mu)}.$$

14. Shape of Body [3]

Free hollow sphere. Inner radius a, outer radius b (see Fig. 9).

Temperature Effect. The heat is generated uniformly inside the material over the whole volume of the sphere. Cooling from the external surface

$$T = \frac{q}{3\lambda}\left[\frac{b^2 - r^2}{2} + \left(\frac{1}{b} - \frac{1}{r}\right)a^3\right].$$

At r = b, T = T_b = 0.

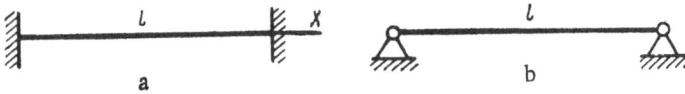

Fig. 10. Various ways of clamping a straight rod:
(a,d) $\mu = 0.5$; (b) $\mu = 1.0$; (c) $\mu = 0.7$.

Stresses

$$\sigma_\Theta = \frac{\alpha q E}{3\lambda(1-\mu)} \left\{ \frac{2r^3 + a^3}{b^3 - a^3} \frac{1}{r^3} \left[\frac{b^3 - a^3}{3} \left(\frac{b^2}{2} + \frac{a^3}{b} \right) - \frac{b^5 - a^5}{10} - \frac{b^2 - a^2}{2} a^3 \right] \right.$$
$$\left. + \frac{1}{r^3} \left[\frac{r^3 - a^3}{3} \left(\frac{b^2}{2} + \frac{a^3}{b} \right) - \frac{r^5 - a^5}{10} - \frac{r^2 - a^2}{2} a^3 \right] - \frac{b^2 - r^2}{2} + \frac{b - r}{br} a^3 \right\};$$

$$\sigma_r = \frac{2\alpha q E}{3\lambda(1-\mu)} \left\{ \frac{r^3 - a^3}{b^3 - a^3} \frac{1}{r^3} \left[\frac{b^3 - a^3}{3} \left(\frac{b^2}{2} + \frac{a^3}{b} \right) - \frac{b^5 - a^5}{10} - \frac{b^2 - a^2}{2} a^3 \right] \right.$$
$$\left. - \frac{1}{r^3} \left[\frac{r^3 - a^3}{3} \left(\frac{b^2}{2} + \frac{a^3}{b} \right) - \frac{r^5 - a^5}{10} - \frac{r^2 - a^2}{2} a^3 \right] \right\}.$$

The maximum value of the peripheral stresses will occur at the surface of the sphere:

At r = a

$$\sigma_\Theta = \frac{\alpha q E}{3\lambda(1-\mu)} \left\{ \frac{3}{b^3 - a^3} \left[\frac{b^3 - a^3}{3} \left(\frac{b^2}{2} + \frac{a^3}{b} \right) - \frac{b^5 - a^5}{10} - \frac{b^2 - a^2}{2} a^3 \right] - \frac{b^2 - a^2}{2} + \frac{b - a}{b} a^2 \right\}.$$

At r = b

$$\sigma_\Theta = \frac{\alpha q E}{3\lambda(1-\mu)} \left[\frac{b^3 - a^3}{3} \left(\frac{b^2}{2} + \frac{a^3}{b} \right) - \frac{b^5 - a^5}{10} - \frac{b^2 - a^2}{2} a^3 \right] \frac{3}{b^3 - a^3}.$$

Rods, Shafts, Beams

15. Shape of Body [6]

Uniform shaft, clamped at the ends (Fig. 10).

Temperature Effect. Uniform heating to T degrees.

Stresses

$$\sigma_x = -\alpha E T.$$

Compression occurs during heating, and tension occurs during cooling. If the compressive stress reaches the critical value (Euler or Yasinskii), bending occurs (loss of strength).

The stresses remain constant with further increase in temperature, while thermal elongation produces longitudinal bending of the shaft. The length reduction factor μ, used in calculating the critical stress, must correspond with supported clamping of the shaft. The coefficient μ is the reciprocal of the number of half-waves in the bent axis of the shaft.

Displacements. There are no displacements before loss of strength occurs.

16. Shape of Body [6]

Long straight rod of any cross section (Fig. 11).

Temperature Effect. The irregular increase in temperature:

$$T = T(x, y, z).$$

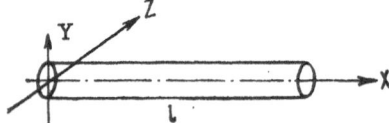

Fig. 11. Straight shaft
of arbitrary cross-sectional shape.

Displacements. The mean absolute increase in the length l of the rod is

$$\Delta l = \frac{1}{F} \int_V \alpha T \, dV,$$

where $V = Fl$ is the volume of the rod, and F is the cross-sectional area.

The angles of inclination of the cross sections (temperature bending) with respect to the principal axes Y, Z of the plane of the cross section are as follows:

Rotation around Z axis:

$$\theta_z = \frac{1}{I_z} \int_V \alpha T y \, dV.$$

Rotation around Y axis:

$$\theta_y = \frac{1}{I_y} \int_V \alpha T z \, dV.$$

There is no thermal torsion, since the rotation of the cross sections around the X axis of the rod is equal to zero.

The components of the linear displacement of one end of the rod with respect to the other along the principal axes of the cross section are as follows:

$$\delta_z = \frac{1}{I_y} \int_V \alpha T z \, (l - x) \, dV,$$

$$\delta_y = \frac{1}{I_z} \int_V \alpha T y \, (l - x) \, dV,$$

where I_z and I_y are the principal moments of inertia of the cross-sectional area of the rod.

17. Shape of Body [8]

Straight rod, clamped at the ends. The rod has a stress concentrator (notch).

Temperature Effect. Rod uniformly heated to T degrees.

Stresses. Far from the notch, the stessed state will be linear and the axial stress will be

$$\sigma_x = -\alpha E T.$$

A stress concentration occurs at the notch, which may be calculated in the same way as for purely mechanical loading of a rod compressed by the force

$$N_x = \alpha E T F.$$

For a plane sample with two cutouts, the dimensions of which are shown in Fig. 12, the stress concentration coefficient may be found from the following approximate formulas [9]:

$$\text{for} \quad a \geqslant \frac{t}{\sqrt{\frac{t}{\rho}}} \quad \alpha_c = \frac{1 + 2\sqrt{\frac{t}{\rho}}}{1 + \frac{t}{a}};$$

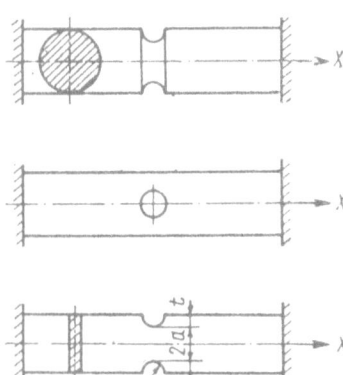

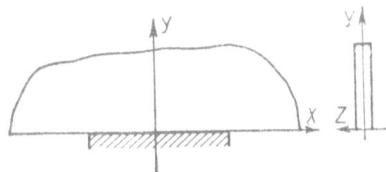

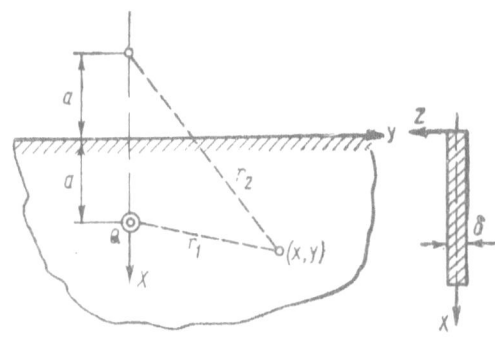

Fig. 12. Notched straight rod.

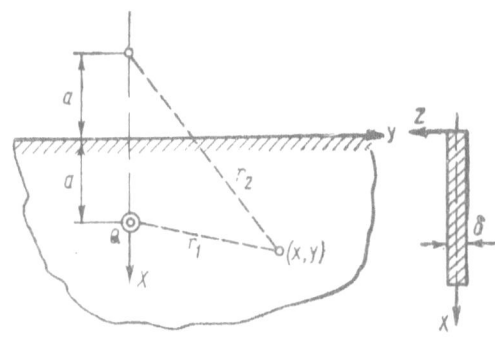

Fig. 13. Semi-infinite plate,
clamped along part of the edge.

Fig. 14. Semi-infinite plate with a point source
of heat.

for $a \leqslant \dfrac{t}{\sqrt{\dfrac{t}{\rho}}}$ $\alpha_c = \dfrac{1 + \dfrac{t}{a} + \sqrt{\dfrac{t}{\rho}}}{1 + \dfrac{t}{a}}$.

At the base of the notch, $\sigma_{max} = \alpha_c \sigma_x$ nom.

Plates, Strips, Sheets, and Half-Planes

18. Shape of Body [10]

Thin, plane, semi-infinite sheet with a straight edge. Clamped for a finite length along the edge (Fig. 13).

Temperature Effect. Uniform elevation of temperature.

Stresses. Infinite stresses occur at the ends of the clamp, and thus under actual conditions, i. e., depending on the degree of departure from ideal conditions, stress concentration occurs at these points. Plastic deformations or flaws may occur at these points.

19. Shape of Body [5]

Half-plane of thickness δ.

Temperature Effect. The temperature rise is produced by the point source Q, located inside the half plane at a distance a from the boundary (Fig. 14). The temperature is assumed to be zero at the boundary. This gives

$$T = \frac{Q}{2\lambda\pi\delta} \ln \frac{r_2}{r_1} ,$$

where

$$r_1 = \sqrt{(x-a)^2 + y^2}, \quad r_2 = \sqrt{(x+a)^2 + y^2}.$$

Stresses

$$\sigma_x = -\frac{\alpha E Q}{4\pi\lambda\delta} \left[\ln \frac{r_2}{r_1} + y^2 \left(\frac{1}{r_2^2} - \frac{1}{r_1^2} \right) \right] + \frac{\alpha Q E a x}{2\pi\lambda\delta} \frac{(x+a)^2 - y^2}{r_2^4} ;$$

$$\sigma_y = -\frac{\alpha E Q}{4\pi\lambda\delta} \left[\ln \frac{r_2}{r_1} + \frac{(x+a)^2}{r_2^2} - \frac{(x-a)^2}{r_1^2} \right] + \frac{\alpha E Q}{4\pi\lambda\delta} \frac{r_2^2 (x+2a) + 2xy^2}{r_2^4} ;$$

$$\tau_{xy} = \frac{\alpha E Q}{4\pi\lambda\delta} \left\{ \left[\frac{x+a}{r_2^2} - \frac{x-a}{r_1^2} \right] y - 2ay \frac{a^2 - x^2 + y^2}{r_2^4} \right\} .$$

20. Shape of Body [6]

Plane, free, thin plate of arbitrary shape in plan (Fig. 15).

Temperature Effect. Steady-state temperature distribution T = T(x,y) satisfying the Laplace equation:

$$\frac{\partial^2 T}{\partial x^2} + \frac{\partial^2 T}{\partial y^2} = 0.$$

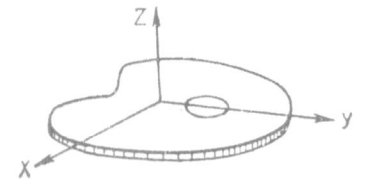

Fig. 15. Free plate of arbitrary shape in plan.

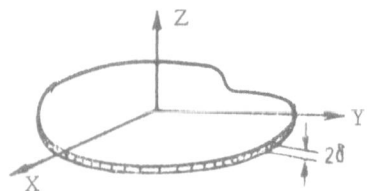

Fig. 16. Plate fastened at the edge.

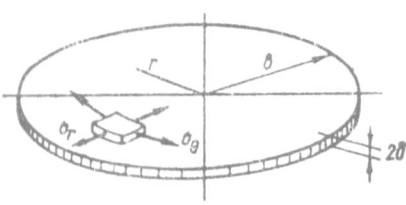

Fig. 17. Plate, circular in plan.

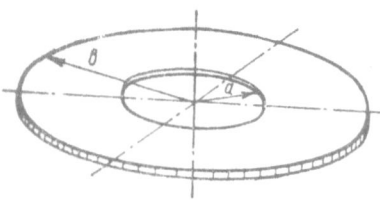

Fig. 18. Circular disc with a hole in the center.

There are no heat sources or sinks inside the region bounded by the outside edge of the plate.

Stresses. No stresses occur.

Deformations

$$\varepsilon_y = \varepsilon_x = \varepsilon_z = \alpha T.$$

21. Shape of Body [6]

Uniform flat plate of arbitrary shape in plan (see Fig. 15). Elongation prevented at all edges (including holes if there are any).

Temperature Effect. Uniform heating to T degrees.

Stresses

$$\sigma_x = \sigma_y = - \frac{\alpha E T}{1 - \mu}.$$

Residual stresses equal to zero.

Deformations

$$\varepsilon_z = \alpha T, \quad \varepsilon_x = \varepsilon_y = 0.$$

22. Shape of Body [6]

Thin flat plate of arbitrary shape in plan, fastened at the edge (Fig. 16).

Temperature Effect. Temperature distribution symmetric about the median plane:

$$T = T(z) = T(-z).$$

Stresses

$$\sigma_x = \sigma_y = \frac{\alpha E}{1 - \mu} \left(\frac{1}{2\delta} \int_{-\delta}^{+\delta} T dz - T \right).$$

23. Shape of Body [5]

Thin flat plate, circular in plan, supported at the edges (Fig. 17).

Temperature Effect. Temperature varies linearly along the thickness alone. One side of the plate is heated, and the other side is cooled by the same amount. Temperature drop between surfaces of the plate, ΔT.

Stresses. Stresses equal to zero.

Displacements. The plate is bent into a spherical surface. The buckling in the Z direction is equal to

$$w = \frac{\alpha \Delta T}{2\delta} \frac{1 + \mu}{1 - \mu} (R^2 - r^2).$$

24. Shape of Body [5]

Flat thin plate (Fig. 18), circular ring in plan (disc with a hole in the center).

Temperature Effect. The constant temperature T is maintained on the outside surface of the hole (r = a). Temperature equal to zero on the outside surface of the ring (r = b). Lateral surface of the disc insulated.

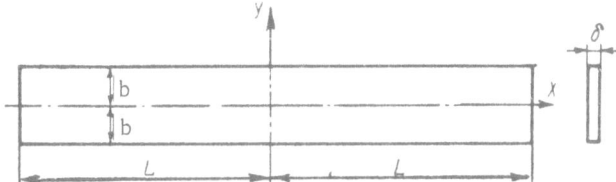

Fig. 19. Long straight thin strip.

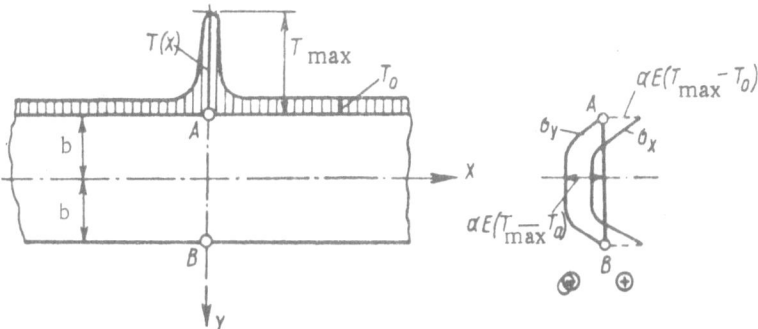

Fig. 20. Temperature and stresses in long thin strip.

Stresses

$$\sigma_\Theta = -\frac{\alpha E T}{2}\left[\frac{\ln\frac{b}{r}-1}{\ln\frac{b}{a}} + \frac{\frac{b^2}{r^2}+1}{\frac{b^2}{a^2}-1}\right];$$

$$\sigma_r = -\frac{\alpha E T}{2}\left[\frac{\ln\frac{b}{r}}{\ln\frac{b}{a}} - \frac{\frac{b^2}{r^2}-1}{\frac{b^2}{a^2}-1}\right];$$

$$\sigma_z = 0.$$

25. Shape of Body [6]

Free thin solid circular disc of radius b (see Fig. 17).

Temperature Effect. Temperature follows an arbitrary law along the radius

$$T = T(r).$$

Stresses

$$\sigma_\Theta = \alpha E\left(\frac{1}{b^2}\int_0^b Trdr + \frac{1}{r^2}\int_0^r Trdr - T\right);$$

$$\sigma_r = \alpha E\left(\frac{1}{b^2}\int_0^b Trdr - \frac{1}{r^2}\int_0^r Trdr\right).$$

Displacements. Radial displacement

$$u = \frac{\alpha(1+\mu)}{r} \int\limits_0^r Tr\,dr + \frac{\alpha(1+\mu)}{b^2} r \int\limits_0^b Tr\,dr.$$

26. Shape of Body [5]

Solid thin circular plate (disc) of constant thickness 2δ (see Fig. 17).

Temperature Effect. Heating produced by a source of strength Q, located in the center of the plate. Lateral surface of plate insulated. Heat removed from edge of disc at r = b, and temperature at edge of disc equal to zero.

$$T = \frac{Q}{4\lambda\pi\delta} \ln \frac{b}{r}.$$

Stresses

$$\sigma_\Theta = -\frac{\alpha E Q}{8\lambda\pi\delta} \left(\ln \frac{b}{r} - 1 \right);$$

$$\sigma_r = -\frac{\alpha E Q}{8\lambda\pi\delta} \ln \frac{b}{r}.$$

In the center, where the concentrated heat source is applied, the stresses approach infinity. Thus, plastic deformations or flaws are to be expected around the center.

Displacements. The radial displacement is equal to

$$u = \frac{\alpha Q r}{8\pi\lambda\delta} \left[(1+\mu) \ln \frac{b}{r} + 1 \right].$$

27. Shape of Body [11, 12]

Long free strip of width 2b and small thickness δ. Here and in what follows, b/L << 1, δ << b (Fig. 19).

Temperature Effect. Temperature varies only from one edge of the strip to the other, T = T(y). Along the strip, the temperature either does not change or changes linearly.

Stresses

$$\sigma_x = \alpha E \left(\frac{1}{2b} \int\limits_{-b}^b T\,dy + \frac{3y}{2b^3} \int\limits_{-b}^b Ty\,dy - T \right);$$

$$\sigma_y = \sigma_z = 0.$$

If bending is prevented (no inclination of the cross sections by suitable clamping), the second component will be missing in the formula. If compression of the strip is prevented (no translational motion of the cross sections), the first component will be lacking. The sign of y must be taken into account in the second component.

If the temperature is linearly dependent on x, in addition to σ_x we have the following tangential stress in the cross section:

$$\tau_{xy} = \alpha E \frac{\partial}{\partial x} \left[\int\limits_{-b}^y T\,dy - \frac{1}{2}\left(1 + \frac{y}{b}\right) \int\limits_{-b}^b T\,dy + \frac{3}{4b}\left(1 - \frac{y^2}{b^2}\right) \int\limits_{-b}^b Ty\,dy \right].$$

The formulas for σ_x and τ_{xy} may also be used in the case where T is not linearly dependent on x. The error will be less, the smaller the ratio b/L.

Fig. 21. Change in the heat supplied
to the end cross section of strip.

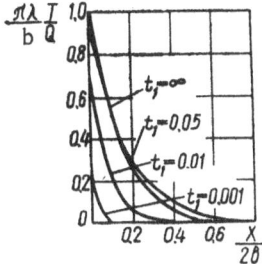

Fig. 22. Change in temperature at end of strip
as a function of time.

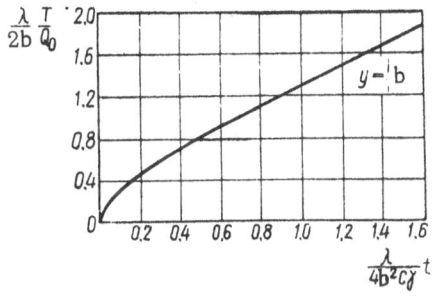

Fig. 23. Increase in temperature with time
on upper edge of strip (y = b).

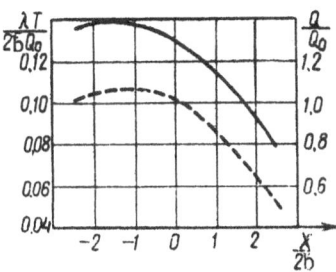

Fig. 24. Change in heat supplied
and temperature along upper edge of strip.

Displacements. The displacement of the cross sections along the Y axis may be found by double integration of the expression for the curvature

$$\frac{d^2v}{dx^2} = \frac{3\alpha}{2b^3} \int_{-b}^{b} Ty\,dy.$$

28. Shape of Body [12]

Long strip (see Fig. 19).

Temperature Effect. T varies along the x axis:

$$T = T(x).$$

Stresses

$$\sigma_x = -\frac{\alpha E}{6}(3y^2 - b^2)\frac{d^2T}{dx^2};$$

$$\sigma_y = -\frac{\alpha E}{24}(y^2 - b^2)\frac{d^4T}{dx^4};$$

$$\tau_{xy} = \frac{\alpha E}{6}(y^2 - b^2)y\frac{d^3T}{dx^3}.$$

Displacements. The displacement of the cross sections v along the Y axis may be found from the equation

$$\frac{d^2v}{dx^2} = -\frac{3\alpha}{20b} \int_{-b}^{b} \frac{d^2T}{dx^2}y\,dy + \frac{\alpha}{4b^3} \int_{-b}^{b} \frac{d^2T}{dx^2}y^3\,dy.$$

29. Shape of Body [6]

Long strip (see Fig. 19). Modulus of elasticity and co-efficient of linear expansion vary over the width:

$$E = E(y), \quad \alpha = \alpha(y).$$

The strip may be made up of horizontal strips of different materials. The thickness δ can vary moderately over the width.

Temperature Effect. The temperature varies only over the width:

$$T = T(y).$$

Stresses

$$\sigma_x = -\alpha ET + c_1Ey + c_2E,$$

where

$$c_1 = \frac{\int_{-b}^{b} \alpha ET\delta y\,dy}{\int_{-b}^{b} E\delta y^2\,dy};$$

35

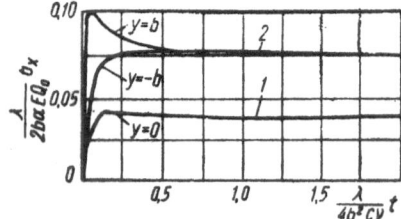

Fig. 25. Stress changes with time on edges and in the middle of strip: (1) tension; (2) compression.

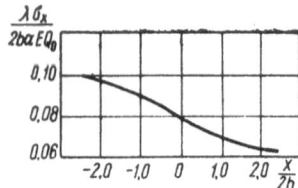

Fig. 26. Stress change along upper edge of strip.

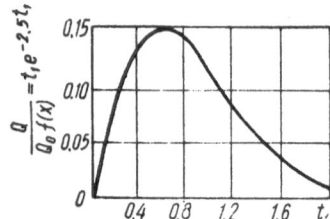

Fig. 27. Change with time in heat added to upper edge of strip.

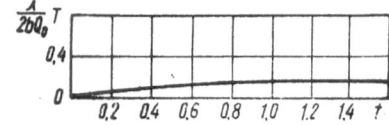

Fig. 28. Temperature change with time on upper edge of strip.

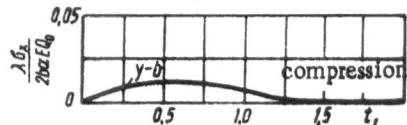

Fig. 29. Stress change with time on lower edge of strip.

$$C_2 = \frac{\int\limits_{-b}^{b} \alpha E T \delta\, dy}{\int\limits_{-b}^{b} E \delta\, dy} \cdot$$

30. Shape of Body [6]

Infinitely long strip of constant thickness δ and width 2b (Fig. 20).

Temperature Effect. Temperature does not change over the width or thickness. A very narrow region of the strip heated along the vertical line AB. T = T(x), but, starting at some distance from the heated line, the temperature may be assumed constant.

Stresses. On the upper and lower edges of the strip, y = ±b, tensile stresses occur as follows:

$$\sigma_x = \alpha E\,(T - T_0).$$

The maximum stresses occur at the points A and B:

$$\sigma_{x\,\max} = \alpha E\,(T_{\max} - T_0).$$

The stresses σ_y will be compressive and the maximum value will be reached at the center of the strip on the line AB:

$$\sigma_{y\,\max} = -\alpha E\,(T_{\max} - T_0).$$

31. Shape of Body [12]

Semi-infinite strip (Fig. 21).

Temperature Effect. At the end of the strip (x = 0) heat begins to be added suddenly (at T = 0) according to the law

$$Q = Q_0 \cos \frac{\pi y}{b},$$

i.e., the central part of the strip is heated while the edges are cooled (for "self-balancing" curve, see Fig. 21).

Saint-Venant's principle holds for the temperature as well as the stresses. The temperature T drops off rapidly with increase in x. It may be seen from Fig. 22 that the temperature increases with the time t, but later on, as the time approaches infinity, the temperature field approaches a steady state, and for values of x greater than the width of the cross section, the temperature is equal to zero. On the graph

$$t_1 = \frac{\lambda}{4b^2\,c\gamma}\,t.$$

Stresses. Like the temperature, stresses only occur where the temperature field is nonuniform (at the end of the strip).

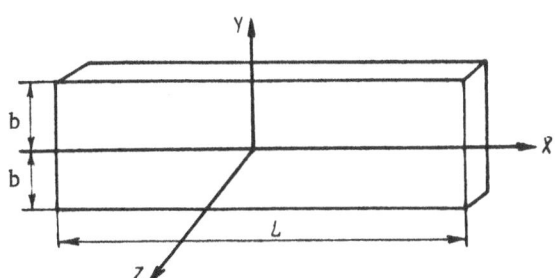

Fig. 30. Long thick strip of rectangular cross section.

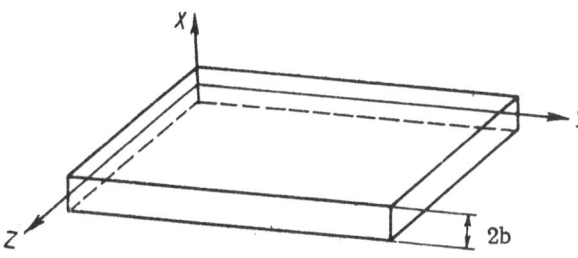

Fig. 31. Thick plate.

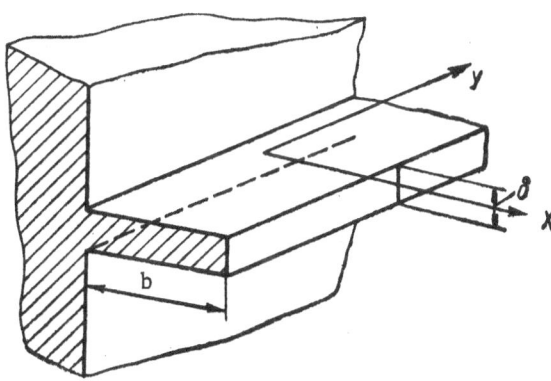

Fig. 32. Plate fastened to a solid.

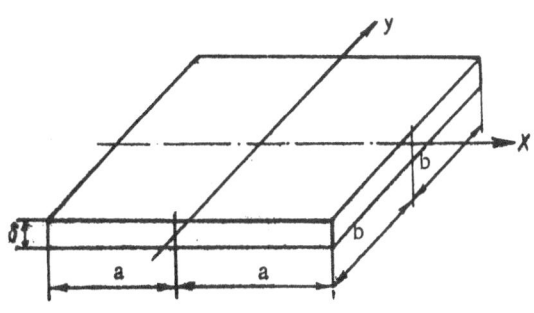

Fig. 33. Rectangular plate.

32. Shape of Body [12]

Long thin strip (see Fig. 19).

<u>Temperature Effect.</u> Heat added to the upper edge of the strip (y = b), the amount depending on x. Lateral surfaces and edges insulated:

$$Q = Q_0 \left(1 - 0.05 \frac{x}{b} - 0.01 \frac{x^2}{b^2} \right).$$

Heat added suddenly and remains constant in time. Temperature increases with time (Fig. 23). Change in temperature T and heat Q along the X axis on the upper edge as shown in Fig. 24 for the instant of time t = 0.2cγ b^2/λ.

<u>Stresses.</u> Figure 25 shows the change with time of the stress σ_x, acting on the cross section x = 0 at the points y = b, y = −b, y = 0.

The change in the stress σ_x on the upper edge of the strip along the X axis (along the strip) for the instant of time t = 4cγ b^2/λ is shown in Fig. 26.

33. Shape of Body [12]

Long thin strip (see Fig. 19). Notation the same as in the preceding case.

<u>Temperature Effect.</u> Heat added to the upper edge of the strip (y = b), and the amount changes along the strip as well as with time, the lateral surfaces and the edges are insulated:

$$Q \doteq Q_0 \left(1 - 0.05 \frac{x}{b} - 0.01 \frac{x^2}{b^2} \right) t_1 e^{-2.5 t_1}.$$

The heat added decreases with time. Here we use the notation

$$t_1 = \frac{\lambda}{4b^2 c\gamma} t$$

and assume b/L = 0.2.

In Fig. 27, the following notation is used for brevity:

$$f(x) = 1 - 0.05 \frac{x}{b} - 0.01 \frac{x^2}{b^2} .$$

The increase in temperature T with time t is shown in Fig. 28 for the upper point at y = b and x = 0. The temperature does not increase significantly with time.

<u>Stresses.</u> Figure 29 shows the change with time t of the stress σ_x on the lower edge of the strip y = −b, at x = 0. The stress decreases to zero with time.

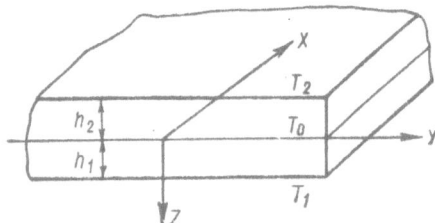

Fig. 34. Plate with a heat-generating layer
at the median plane.

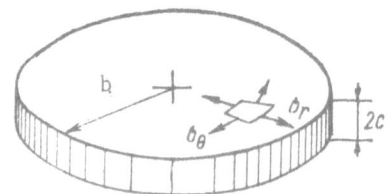

Fig. 35. Round plate.

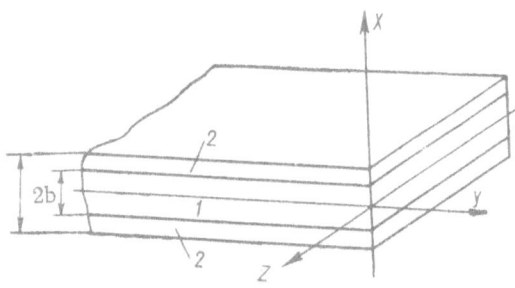

Fig. 36. Three-layer compound plate,
symmetric about the median plane.

34. Shape of Body [8]

Long thick free strip of width 2b (Fig. 30).

Temperature Effect. The heat is uniformly generated inside the material of the strip. Heat is removed symmetrically from the upper and lower sides of the strip at y = ±b. The temperature on the surface is taken equal to zero.

$$T = \frac{q}{2\lambda}(b^2 - y^2).$$

Stresses

$$\sigma_x = \frac{\alpha q E}{2\lambda(1-\mu)}\left(y^2 - \frac{b^2}{3}\right).$$

The stress becomes maximum on the surfaces of the strip at y = ±b,

$$\sigma_{x\,max} = \frac{\alpha q E b^2}{3\lambda(1-\mu)}.$$

35. Shape of Body [3]

Free flat thick plate (Fig. 31).

Temperature Effect. The temperature varies only over the thickness:

$$T = T(x).$$

Stresses. Far from the free edges, the stresses will be:

$$\sigma_z = \sigma_y = \frac{\alpha E}{1-\mu}\left(\frac{1}{2b}\int_{-b}^{b} T dx + \frac{3x}{2b^3}\int_{-b}^{b} Tx\,dx - T\right).$$

36. Shape of Body [5]

Long strip fastened to a solid (cooling rib, Fig. 32).

Temperature Effect. Temperature difference between the solid being cooled (at the base of the rib) and the cooling medium equal to ΔT.

Stresses

$$\sigma_y = -\alpha E \Delta T \frac{\operatorname{ch} m(b-x)}{\operatorname{ch} mb},$$

where $m = \sqrt{2k/\lambda\delta}$, and k is the heat transfer coefficient from the surface of the rib. Maximum stresses at the base of the rib, at x = 0:

$$\sigma_{y\,max} = -\alpha E \Delta T.$$

37. Shape of Body [5]

Rectangular plate, a ≤ b (Fig. 33).

Temperature Effect. The upper and lower surfaces of the plate are kept at the same temperature, but of different signs.

On the upper surface: $T = T(x, y).$

TABLE 4. Values of the Coefficients c_x and c_y

b/a	1	1.5	2	2.5	3
c_x	0.651	0.843	0.914	0.963	0.99
c_y	0.651	0.465	0.405	0.342	0.33

On the lower surface: $T = -T(x, y)$.

Stresses. Edges of the plate clamped. The maximum stresses will be at the surfaces:

$$\sigma_x = \sigma_y = -\frac{\alpha E T}{1 - \mu},$$

on the upper surface — compression, on the lower surface — tension.

Edges of the plate supported, but the temperature on the surface independent of x, y:

$$T = \text{const} = \pm T.$$

The stresses on the surface of the plate are

$$\sigma_x = \frac{\alpha E T}{1 - \mu}(c_x - 1); \quad \sigma_y = \frac{\alpha E T}{1 - \mu}(c_y - 1);$$

the coefficients c_x and c_y are given in Table 4. Here the X axis is directed along the short side of the plate (in the direction 2a).

38. Shape of Body [5]

Plate of thickness $H = h_1 + h_2$ (Fig. 34). Lateral surfaces of the plate: $x = \pm a$, $y = \pm b$.

Temperature Effect. The heat-generating layer is located in the plane XY. The temperature of the layer is T_0, the temperatures of the upper and lower surfaces of the plate are T_1 at $z = h_1$ and T_2 at $z = -h_2$. T_0, T_1, and T_2 are independent of x, y, and t. Here the temperature will be linearly distributed along the height.

Stresses. Edges of the plate are clamped, but can move in the direction x, y:

for $z = h_1$,

$$\sigma_x = \sigma_y = -\frac{\alpha E}{2(1 - \mu)}\left[\frac{h_2}{H}(T_1 - T_2) - T_0 + T_1\right];$$

for $z = 0$,

$$\sigma_x = \sigma_y = -\frac{\alpha E}{2(1 - \mu)}\left[\frac{h_1}{H}(T_0 - T_1) + \frac{h_2}{H}(T_0 - T_2)\right];$$

for $z = -h_2$,

$$\sigma_x = \sigma_y = -\frac{\alpha E}{2(1 - \mu)}\left[T_2 - T_0 - \frac{h_1}{H}(T_1 + T_0)\right].$$

Displacements. The displacements along the Z axis are equal to zero.

39. Shape of Body [6]

Solid flat circular plate of radius b, and thickness 2c (Fig. 35).

Temperature Effect. The upper surface is heated and the lower surface is cooled to the same temperature. T is the temperature drop between the surfaces of the plate. Temperature varies linearly over the thickness, and arbitrarily along the radius:

$$T = T(r).$$

1. Plate fastened around the edge.

Stresses

$$\sigma_\theta = \frac{\alpha E}{2} \left(T - \frac{1}{r^2} \int_0^r Tr\,dr + \frac{1+\mu}{1-\mu} \frac{1}{b^2} \int_0^b Tr\,dr \right);$$

$$\sigma_r = \frac{\alpha E}{2} \left(\frac{1}{r^2} \int_0^r Tr\,dr + \frac{1+\mu}{1-\mu} \frac{1}{b^2} \int_0^b Tr\,dr \right).$$

Displacements. Perpendicular to the median surface

$$w = \frac{1+\mu}{2c} \left[\int_0^r \frac{F(r)}{r}\,dr - \int_0^b \frac{F(r)}{r}\,dr + \frac{F(b)}{2} \left(1 - \frac{r^2}{b^2} \right) \right],$$

where

$$F(r) = \int_0^r \alpha Tr\,dr.$$

2. Plate supported around the edge.

Stresses

$$\sigma_\theta = \frac{\alpha E}{2} \left(T - \frac{1}{r^2} \int_0^r Tr\,dr - \frac{1}{b^2} \int_0^b Tr\,dr \right);$$

$$\sigma_r = \frac{\alpha E}{2} \left(\frac{1}{r^2} \int_0^r Tr\,dr - \frac{1}{b^2} \int_0^b Tr\,dr \right).$$

Displacements. Perpendicular to the median plane:

$$w = \frac{1+\mu}{2c} \left[\int_0^r \frac{F(r)}{r}\,dr + \frac{1-\mu}{1+\mu} \frac{F(b)}{2} \left(\frac{r^2}{b^2} - 1 \right) - \int_0^b \frac{F(r)}{r}\,dr \right].$$

40. Shape of Body [13]

Flat plate 1, tightly covered on two sides by slabs of another material 2 (Fig. 36).

Temperature Effect. Heat generated in layer 1. Temperature changes only along the thickness:

$$T = T(x).$$

Stresses. A plane stressed state occurs, $\sigma_z = \sigma_y = \sigma$.

Stress in layer 1:

$$\sigma = -\xi_1 T_1 + \frac{3}{2} x \xi_{12} \frac{\int_{-a}^b \xi_2 x T_2\,dx + \int_{-b}^b \xi_1 x T_1\,dx + \int_b^a \xi_2 x T_2\,dx}{a^3 + b^3(\xi_{12} - 1)} + \frac{1}{2} \xi_{12} \frac{\int_{-a}^{-b} \xi_2 T_2\,dx + \int_{-b}^b \xi_1 T_1\,dx + \int_b^a \xi_2 T_2\,dx}{a + b(\xi_{12} - 1)}.$$

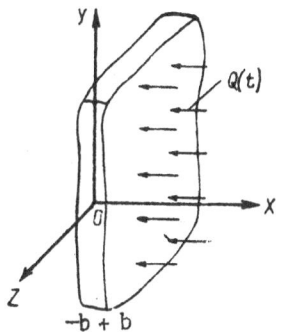

Stress in layer 2:

$$\sigma = -\,\xi_2 T_2 + \frac{3}{2}\,x\,\frac{\displaystyle\int_{-a}^{-b}\xi_2 x T_2\,dx + \int_{-b}^{b}\xi_1 x T_1\,dx + \int_{b}^{a}\xi_2 x T_2\,dx}{a^3 + b^3(\xi_{12}-1)}$$

$$+\,\frac{1}{2}\,\frac{\displaystyle\int_{-a}^{-b}\xi_2 T_2\,dx + \int_{-b}^{b}\xi_1 T_1\,dx + \int_{b}^{a}\xi_2 T_2\,dx}{a + b(\xi_{12}-1)}\,,$$

where

Fig. 37. Heat supplied to the surface of a plate.

$$\xi_1 = \frac{\alpha_1 E_1}{1 - \mu_1};\quad \xi_2 = \frac{\alpha_2 E_2}{1 - \mu_2};\quad \xi_{12} = \frac{E_1}{E_2}\,\frac{1 - \mu_2}{1 - \mu_1}.$$

41. Shape of Body

Thick free flat plate (see Fig. 31).

Temperature Effect. Heat uniformly generated inside the material of the plate.

The plate is cooled on both the lateral surfaces, and ΔT is the temperature drop between the surfaces. Maximum temperature at the distance $x_0 = \lambda \Delta T / q$ from the center of the plate

$$T_{\mathrm{max}} = \frac{q\delta^2}{2\lambda}\left(1 + \frac{\lambda \Delta T}{2q\delta^2}\right)^2.$$

Stress. Maximum stress, occurring at the surface

$$\sigma_z = \sigma_y = \frac{\alpha q E \delta^2}{3\lambda(1 - \mu)} + \frac{\alpha E}{1 - \mu}\,\frac{\Delta T}{2}.$$

42. Shape of Body [14]

Flat free plate of constant thickness 2b, and arbitrary shape in the YZ plane.

Temperature Effect. One surface of the plate and all the edges are insulated. Heat supplied uniformly to the other surface of the plate (Fig. 37). Heat supply time dependent, $Q = Q(t)$ [cal/cm²-sec]. Heat flux increases from zero to some maximum value Q_M in time t_M and then gradually decreases to zero.

At the initial instant, $Q(0) = 0$.

In this case, the temperature distribution in the plate will be of the form

$$T(x, t) = \frac{Q(t)}{\lambda}\left(\frac{x^2}{4b} + \frac{x}{2} - \frac{b}{12}\right) + \frac{1}{2b\gamma c}\int_{0}^{t} Q(t)\,dt.$$

Stresses. The stressed state is symmetric about the median plane:

$$\sigma_y = \sigma_z = \sigma;\quad \sigma_x = \tau_{xy} = \tau_{xz} = \tau_{yz} = 0.$$

Operation of a Plate in the Elastic Stage (see Table 5)

1. Stresses at points in the plate will be:

$$\sigma = \frac{\alpha E Q}{(1 - \mu)\cdot 4b\lambda}\left(\frac{b^2}{3} - x^2\right).$$

The stresses σ reach a maximum compressive value on the surface of the plate and are equal to

$$\sigma = -\,\frac{\alpha Q E b}{6(1 - \mu)\lambda}.$$

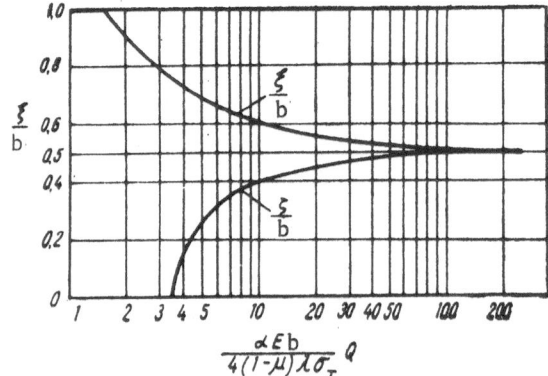

Fig. 38. Change in the boundaries of elastic and plastic zones as the amount of heat supplied to the plate is increased.

In the center of the plate, the maximum tensile stresses will be

$$\sigma = \frac{\alpha Q E b}{12(1-\mu)\lambda}.$$

Operation of a Plate in the Elastoplastic Stage

2. Two plastic regions occur in the plate. They are formed on each surface of the plate with increase in the heat supplied Q.[*]

The plastic regions occur at the instant in which

$$Q = Q_1 = \frac{6(1-\mu)\sigma_T\lambda}{\alpha E b},$$

or when the surfaces reach the flow limit on the surface.

If the heat supplied continues to increase ($Q > Q_1$), the two plastic regions will begin to approach one another. These regions will be located symmetrically about the median plane, i.e., at the following values of x:

$$\xi \leqslant x \leqslant b \text{ and } -b \leqslant x \leqslant -\xi.$$

The stresses in the elastic region (for $|x| < \xi$) will be:

$$\sigma = \frac{\alpha Q E}{4(1-\mu)\lambda b}\left(\frac{\xi^2}{3}-x^2\right) + \frac{b-\xi}{\xi}\sigma_T;$$

and in the plastic region (for $|x| > \xi$)

$$\sigma = -\sigma_T.$$

The way in which the position of the plastic regions varies with the amount of heat supplied is given by the following formula:

$$\xi = b\sqrt[3]{\frac{6(1-\mu)\lambda\sigma_T}{\alpha Q E b}}.$$

3. If the amount of heat supplied continues to increase, a plastic zone will also occur in the central part of the plate when the amount of heat supplied reaches the value

$$Q_2 = \frac{128}{9}\frac{(1-\mu)\sigma_T\lambda}{\alpha E b}.$$

Here $\xi = \frac{3}{4}b$.

If there are three plastic zones in the plate for which x varies within the limits $0 \leq |x| \leq \zeta$ for the central zone and $\xi \leq |x| \leq b$ for the zones on the surface of the plate, the stresses will be:

In the elastic range ($\zeta \leq x \leq \xi$)

$$\sigma = \frac{\alpha Q E}{4(1-\mu)\lambda b}\left[\frac{\xi^3-\zeta^3}{3(\xi-\zeta)}-x^2\right] + \frac{b-\xi-\zeta}{\xi-\zeta}\sigma_T.$$

----
[*]See Table 5.

In the plastic ranges ($-\zeta \le x \le \zeta$; $\xi \le |x| \le b$)

$$\sigma = \pm \, \sigma_T.$$

The value of ζ may be found from the equation

$$\xi + \zeta = \frac{8\,(1-\mu)\,\sigma_T\,\lambda}{\alpha Q E \,(\xi - \zeta)\,b^3}.$$

Figure 38 shows how ξ and ζ change with increase in the amount of heat supplied.

Residual Stresses

1. If the heat flux begins to decrease after the maximum value Q_{max}, without reaching the value Q_1, no residual stresses occur, since the plate does not leave the elastic range.[*]

2. In the case $Q_1 < Q_M < Q_a$, where

$$Q_a = \frac{12\,(1-\mu)\,\sigma_T\,\lambda}{\alpha E b},$$

the plate has two plastic zones (at the surfaces). After the heat flux stops, the residual stresses will be:

for $0 \le |x| \le \xi_M$

$$\sigma = \frac{\alpha Q_M E}{12\,(1-\mu)\,\lambda b}\,(\xi_M^2 - b^2) + \frac{b-\xi_M}{\xi_M}\,\sigma_T;$$

for $\xi_M \le |x| \le b$

$$\sigma = \frac{\alpha Q_M E}{4\,(1-\mu)\,\lambda b}\left(x^2 - \frac{b^2}{3}\right) - \sigma_T,$$

where

$$\xi_M = b\,\sqrt[3]{\frac{6\,(1-\mu)\,\sigma_T\,\lambda}{\alpha Q_M E b}}.$$

The maximum residual stresses will be at the surface, for $x = \pm b$.

3. If

$$\frac{12\,(1-\mu)\,\sigma_T\,\lambda}{\alpha E b} < Q_M < \frac{128\,(1-\mu)\,\sigma_T\,\lambda}{9\alpha E b},$$

the residual stresses at the surface will reach the flow limit, and are found as follows:

for $0 \le |x| \le \xi_M$

$$\sigma = \frac{\alpha Q_M E}{12\,(1-\mu)\,\lambda b}\,(\xi_M^2 - \delta^2) + \frac{b-\xi_M}{\xi_M}\,\sigma_T - 2\,\frac{b-\delta}{\delta}\,\sigma_T;$$

for $\xi_M \le |x| \le \delta$

$$\sigma = \frac{\alpha Q_M E}{4\,(1-\mu)\,\lambda b}\left(x^2 - \frac{\delta^2}{3}\right) - 2\,\frac{b-\delta}{\delta}\,\sigma_T - \sigma_T;$$

for $\delta \le |x| \le b$

$$\sigma = \sigma_T,$$

where

$$\delta = b\,\sqrt[3]{\frac{12\,(1-\mu)\,\sigma_T\,\lambda}{\alpha Q_M E b}}.$$

---

[*]See Table 5.

4. If

$$\frac{128\,(1-\mu)\,\sigma_T\,\lambda}{9aEb} < Q_\text{M} < \frac{256\,(1-\mu)\,\sigma_T\,\lambda}{9aEb},$$

the residual stresses are found thus:

for $0 \leq |\,x\,| \leq \zeta_\text{M}$

$$\sigma = \sigma_T - \frac{aQ_\text{M}E}{4\,(1-\mu)\,\lambda b}\left(\frac{\delta^2}{3} - x^2\right) - 2\frac{b-\delta}{\delta}\,\sigma_T;$$

for $\zeta_\text{M} \leq |\,x\,| \leq \xi_\text{M}$

$$\sigma = \frac{aQ_\text{M}E}{12\,(1-\mu)\,\lambda b}\left(\frac{\xi_\text{M}^3 - \zeta_\text{M}^3}{\xi_\text{M} - \zeta_\text{M}} - \delta^2\right) + \frac{b - \xi_\text{M} - \zeta_\text{M}}{\xi_\text{M} - \zeta_\text{M}}\,\sigma_T - 2\frac{b-\delta}{\delta}\,\sigma_T;$$

for $\xi_\text{M} \leq |\,x\,| \leq \delta$

$$\sigma = \frac{aQ_\text{M}E}{.\,4\,(1-\mu)\,\lambda b}\left(x^2 - \frac{\delta^2}{3}\right) - 2\frac{b-\delta}{\delta}\,\sigma_T - \sigma_T;$$

for $\delta \leq |\,x\,| \leq b$

$$\sigma = \sigma_T.$$

5. If

$$\frac{256\,(1-\mu)\,\sigma_T\,\lambda}{9aEb} < Q_\text{M} < \frac{52\,(1-\mu)\,\sigma_T\,\lambda}{aEb},$$

the residual stresses have three plastic ranges and are found in the following way:

for $0 \leq |x| \leq \eta$

$$\sigma = -\sigma_T;$$

for $\eta \leq |x| \leq \zeta_\text{M}$

$$\sigma = \sigma_T - \frac{aQ_\text{M}E}{4\,(1-\mu)\,\lambda b}\left[\frac{\delta^3 - \eta^3}{3\,(\delta - \eta)} - x^2\right] - 2\frac{b - \delta - \eta}{\delta - \eta}\,\sigma_T;$$

for $\zeta_\text{M} \leq |\,x\,| \leq \xi_\text{M}$

$$\sigma = \frac{aQ_\text{M}E}{12\,(1-\mu)\,\lambda b}\left[\frac{\xi_\text{M}^3 - \zeta_\text{M}^3}{\xi_\text{M} - \zeta_\text{M}} - \frac{\delta^3 - \eta^3}{\delta - \eta}\right] + \frac{b - \xi_\text{M} - \zeta_\text{M}}{\xi_\text{M} - \zeta_\text{M}}\,\sigma_T - 2\frac{b-\delta-\eta}{\delta - \eta}\,\sigma_T;$$

for $\xi_\text{M} \leq |x| \leq \delta$

$$\sigma = \frac{aQ_\text{M}E}{4\,(1-\mu)\,\lambda b^2}\left[x - \frac{\delta^3 - \eta^3}{3\,(\delta - \eta)}\right] - 2\frac{b - \delta - \eta}{\delta - \eta}\,\sigma_T - \sigma_T;$$

for $\delta \leq |x| \leq b$

$$\sigma = \sigma_T,$$

where

$$\delta\,(Q_\text{M}) = \xi\left(\frac{Q_\text{M}}{2}\right);$$

$$\eta\,(Q_\text{M}) = \zeta\left(\frac{Q_\text{M}}{2}\right);$$

$$0 \leqslant \eta \leqslant \zeta_\text{M} \leqslant \xi_\text{M} \leqslant \delta < b.$$

TABLE 5. Stress Curves Along Thickness of Plate

	Acting stresses			Residual stresses	
	Case calc.	Curve of σ	Case calc.	Curve of σ
$0 - \frac{2}{2}$	1	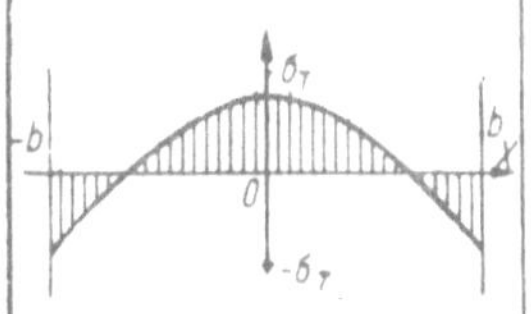	1	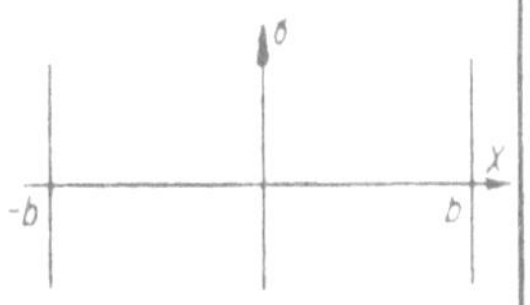
$\frac{3}{2} - 3$			2	
$3 - \frac{32}{9}$	2		3	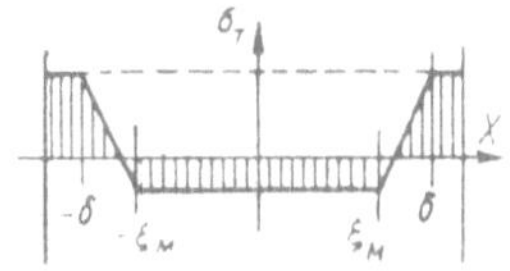
$\frac{32}{5} - \frac{64}{9}$		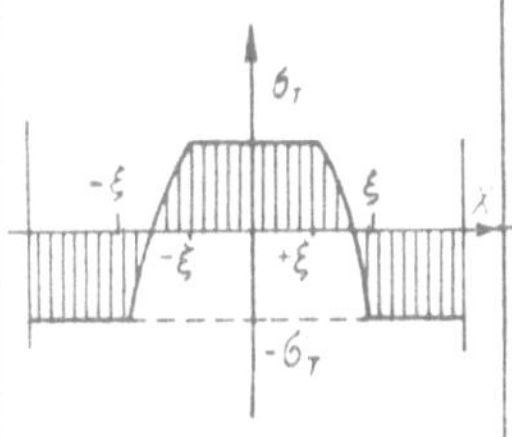	4	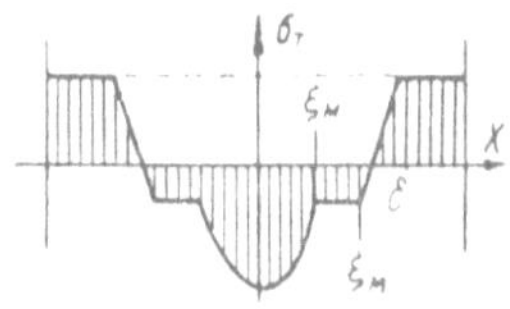
$\frac{64}{9} - 13$	3		5	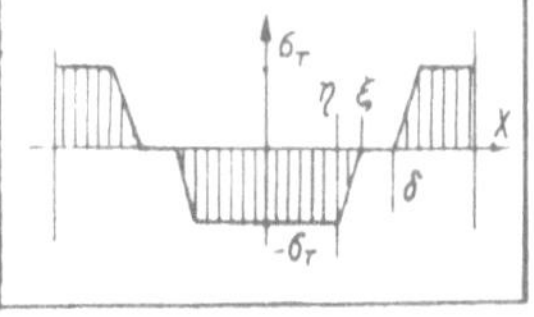

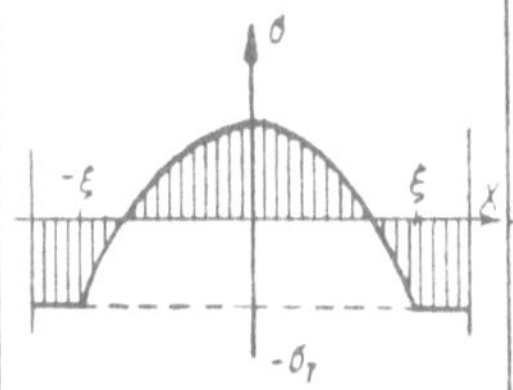

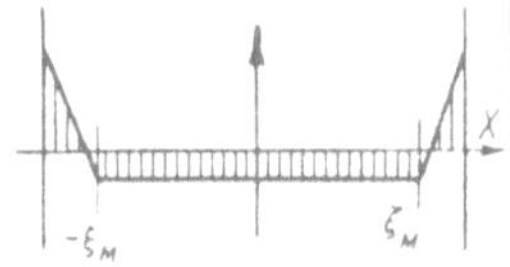

For

$$Q_{\text{M}} > \frac{52(1-\mu)\,\sigma_T\,\lambda}{\pi E b}$$

the plastic regions almost completely fill the thickness of the plate both when the thermal load is acting and after it is taken off. This exhausts the bearing power of the plate.

Table 5 shows plots of the stresses $\sigma = \sigma_y = \sigma_z$, produced by a given thermal load, and the corresponding residual stresses after the heat load is removed.

43. Shape of Body [3]

Flat plate of thickness 2b (or of small curvature in comparison with the thickness), made of linearly viscoelastic material.

Temperature Effect. Transient heating of both surfaces of the plate symmetric about the median plane (or for a plate of half the thickness, heating on one side and insulation on the other, see Fig. 37). The surface temperature is changing rapidly (thermal shock) according to the law

$$\frac{T_{\text{m}} - T_s}{T_0} = -A\left(\frac{t}{\Theta}\right)^{1/4} e^{-at/\Theta},$$

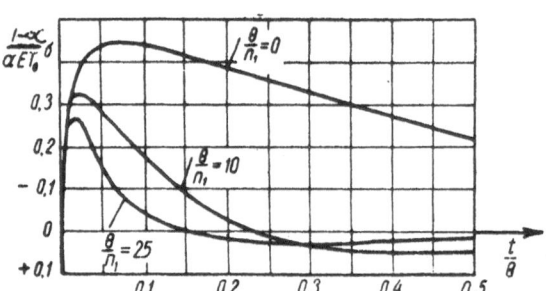

Fig. 39. Change with time in the stresses
on the surface of an elastoviscous plate.

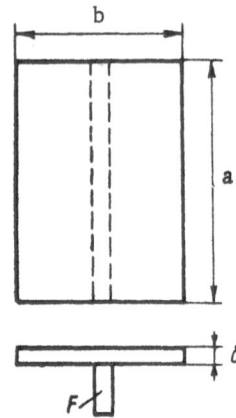

Fig. 40. Piece of siding
with stringer.

where T_m is the mean temperature over the thickness of the plate, T_s is the temperature at the heated surface of the plate, and T_0 is the peak change in ambient temperature above the initial value.

Stresses. The stresses at the surface ($\sigma_y = \sigma_z = \sigma$, $\sigma_x = 0$) vary and attenuate with time according to the law

$$\sigma = \frac{aE(T_m - T_s)}{1-\mu} \left\{ 1 - \frac{4}{15} \frac{1+\mu}{1-\mu} \frac{t}{n} \left[1 + \frac{1}{9} \left(\frac{4}{3} \frac{1+\mu}{1-\mu} \frac{t}{n} - 1 \right) \right] \right\},$$

where $n = k/E$ is the relaxation time of the material and k [kg-sec/cm²] is the coefficient of viscosity or internal resistance coefficient, proportional to the deformation rate. Figure 39 shows the change with time in the stresses on the surface of the plate with A = 1.17, a = 3.1 (Θ is the time parameter), and $n_1 = 3n(1 - \mu)/(1 + \mu)$.

The upper curve ($\Theta/n_1 = 0$) is for an elastic material, and is also the change in temperature with time, $T_m - T_s$.

44. Shape of Body [11]

Idealized piece of siding with a stringer (Fig. 40).

Temperature Effect. Uniform heating of the siding to T_0, and of the stringer to T_c degrees with respect to the original temperature, which is the same for both parts.

Stresses. The stresses are uniformly distributed in siding and stringer:

In siding

$$\sigma_0 = - H \left(\frac{a_0}{a_c} T_0 - T_c \right).$$

In stringer

$$\sigma_c = - \frac{b\delta}{F} \sigma_0,$$

where

$$H = \frac{a_c E_0}{1 + \dfrac{b\delta E_0}{FE_c}} \; ;$$

F is the cross-sectional area of the stringer and α_0, α_c, E_0, and E_c are the coefficients of linear expansion and the moduli of elasticity of the siding and stringer, respectively.

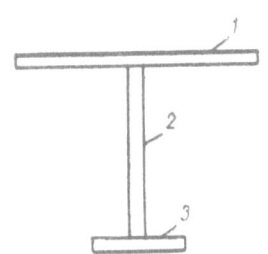

Fig. 41. Piece of siding with stringer: (1) siding; (2) wall; (3) shelf.

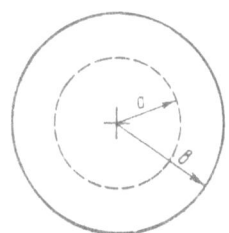

Fig. 42. Hollow sphere.

Fig. 43. Thin-walled straight tube of arbitrary cross section.

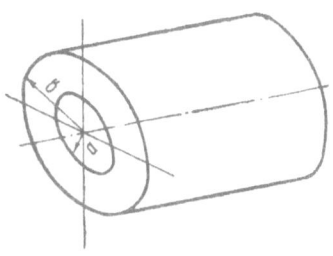

Fig. 44. Long thin-walled tube.

45. Shape of Body [11]

Idealized piece of siding with stringer (Fig. 41). The wall and shelf are made of the same material.

Temperature Effect. The siding is uniformly heated by the amount T_0 with respect to the original uniform temperature, while the wall and the shelf are heated to T_c degrees.

Stresses. The uniform stress in the siding will be:

$$\sigma_0 = - H \left(\frac{\alpha_0}{\alpha_c} T_0 - T_c \right),$$

where

$$H = \frac{\alpha_c E_0 \left(1 + \frac{4 E_c F_s}{E_c F_c} \right)}{1 + \frac{4 E_0 F_0}{E_c F_c} + \frac{4 E_c F_s}{E_c F_c} + \frac{12 E_0 F_0 F_s}{E_c F_c^2}} ;$$

and F_0, F_c, and F_s are the cross-sectional areas of the siding, the wall, and the shelf. The remaining notation is the same as in paragraph 44.

Cylinders and Tubes

46. Shape of Body [6]

Closed hollow sphere. Inner radius a, outer radius b (Fig. 42).

Temperature Effect. Nonuniform temperature rise:

$$T = T(x, y, z).$$

Volume Deformation. The volume $\frac{4}{3}\pi a^3$ occupied by the cavity increases by the amount

$$\Delta V_a = \frac{a^3}{b^3 - a^3} \int_{r=a}^{r=b} 3\alpha T dV.$$

The external volume, $\frac{4}{3}\pi b^3$, occupied by the envelope, is increased by the amount

$$\Delta V_b = \frac{b^3}{b^3 - a^3} \int_{r=a}^{r=b} 3\alpha T dV,$$

where the integrals are taken over the volume of the envelope.

47. Shape of Body [6]

Long straight thin-walled tube of arbitrary cross section (Fig. 43).

Temperature Effect. The temperature of the inner surface of the tube is T_1, and of the outer surface T_2. The temperature varies linearly over the thickness.

Stresses. The peripheral stresses on the inner surface are

$$\sigma_\theta = - \frac{\alpha E}{1 - \mu} \frac{T_1 - T_2}{2},$$

and on the outer surface

$$\sigma_\theta = \frac{\alpha E}{1 - \mu} \frac{T_1 - T_2}{2}.$$

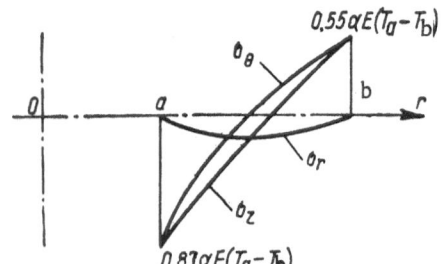

Fig. 45. Stress variation over thickness of tube wall for steady heat flux from inside to outside.

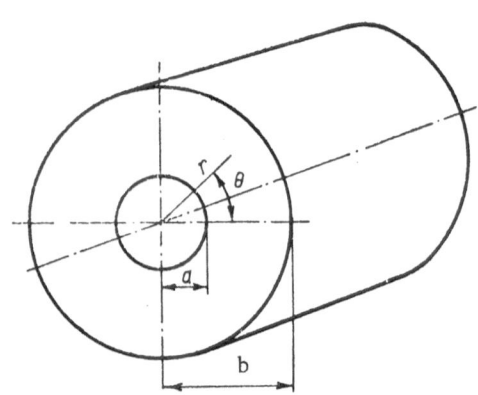

Fig. 46. Long thick-walled cylinder.

48. Shape of Body [5]

Long thin-walled cylinder (Fig. 44).

Temperature Effect. The temperature follows an arbitrary law over the thickness and is symmetric about the axis of the tube:

$$T = T(r).$$

Stresses

$$\sigma_\Theta = \frac{\alpha E}{1-\mu} \left[\left(1 + \frac{a^2}{r^2} \right) \frac{1}{b^2 - a^2} \int_a^b Tr\,dr + \frac{1}{r^2} \int_a^r Tr\,dr - T \right];$$

$$\sigma_r = \frac{\alpha E}{1-\mu} \left[\left(1 - \frac{a^2}{r^2} \right) \frac{1}{b^2 - a^2} \int_a^b Tr\,dr - \frac{1}{r^2} \int_a^r Tr\,dr \right].$$

The radial stress is equal to zero on the inner and outer surfaces of the tube. The axial stress, σ_z, will be different, depending on the way the tubes are clamped at the ends, but this difference has no effect on the values of σ_Θ and σ_r at a distance greater than one diameter from the end of the tube.

If the ends of the tube are free (axial force equal to zero), the axial deformation ε_z = const $\neq$ 0.

$$\sigma_z = \frac{\alpha E}{1-\mu} \left[\frac{2}{b^2 - a^2} \int_a^b Tr\,dr - T \right],$$

or

$$\sigma_z = \sigma_\Theta + \sigma_r.$$

The axial and ring stresses are equal to one another on the inner and outer surfaces of the tube.

With the ends rigidly clamped, the axial deformation ε_z = 0. In this case,

$$\sigma_z = \mu \left(\sigma_\Theta + \sigma_r \right) - \alpha ET.$$

The case of a solid cylinder may be found from this, by setting a = 0.

Displacements. The radial displacement is the same for zero longitudinal force and zero axial deformation:

$$u = \frac{\alpha r}{1-\mu} \left\{ \frac{1}{b^2 - a^2} \left[1 - 3\mu + (1+\mu)\frac{a^2}{r^2} \right] \times \int_a^b Tr\,dr + \frac{1+\mu}{r^2} \int_a^r Tr\,dr \right\}.$$

Displacements on the surfaces of the tube:

for r = b

$$u = \frac{2\alpha b}{b^2 - a^2} \int_a^b Tr\,dr,$$

for r = a

$$u = \frac{2\alpha a}{b^2 - a^2} \int\limits_a^b Tr\,dr.$$

Deformation. With no axial force, i.e., with $\int\limits_a^b \sigma_z \cdot 2\pi\, r\,dr = 0$, the axial deformation is constant for

any cross section of the tube and is independent of the radius:

$$\varepsilon_z = \frac{2}{b^2 - a^2} \int\limits_a^b Tr\,dr,$$

$$\varepsilon_\theta = \frac{u}{r}, \quad \varepsilon_r = \frac{du}{dr}.$$

49. Shape of Body [5]

Long thick-walled cylinder (see Fig. 44), c = b/a, where a is the inner, and b is the outer radius of the tube.

Temperature Effect. Steady heat flux through the wall of the tube. The temperature is T_a on the inner surface, and T_b on the outer surface.

$$T = (T_a - T_b)\,\frac{\ln \dfrac{b}{r}}{\ln c}.$$

Stresses

$$\sigma_\theta = \frac{\alpha E\,(T_a - T_b)}{2\,(1 - \mu)} \left(\frac{1 - \ln \dfrac{b}{r}}{\ln c} - \frac{\dfrac{b^2}{r^2} + 1}{c^2 - 1} \right);$$

$$\sigma_r = \frac{\alpha E\,(T_a - T_b)}{2\,(1 - \mu)} \left(\frac{\dfrac{b^2}{r^2} - 1}{c^2 - 1} - \frac{\ln \dfrac{b}{r}}{\ln c} \right).$$

The axial stress for zero axial force is

$$\sigma_z = \frac{\alpha E\,(T_a - T_b)}{2\,(1 - \mu)} \left(\frac{1 - 2 \ln \dfrac{b}{r}}{\ln c} - \frac{2}{c^2 - 1} \right).$$

For zero axial deformation

$$\sigma_z = \frac{\alpha E\,(T_a - T_b)}{2\,(1 - \mu)} \left(\frac{\mu - 2 \ln \dfrac{b}{r}}{\ln c} + \frac{2\mu}{c^2 - 1} \right).$$

Figure 45 shows the stress curves for c = 2 (σ_z for zero axial force).

The maximum values of the stresses σ_θ and σ_z will occur on the inner and outer surfaces:

for r = a (zero axial stress)

$$\sigma_\theta = \sigma_z = \frac{\alpha E\,(T_a - T_b)}{2\,(1 - \mu)} \left(\frac{1}{\ln c} - \frac{2c^2}{c^2 - 1} \right);$$

for r = b

$$\sigma_\Theta = \sigma_z = \frac{\alpha E (T_a - T_b)}{2 (1 - \mu)} \left(\frac{1}{\ln c} - \frac{2}{c^2 - 1} \right).$$

When the ratio of the radii is small ($1 < c \leq 2$), the stresses on the tube surfaces may be calculated from the following formulas:

for r = a

$$\sigma_\Theta = \sigma_z = \frac{\alpha E (T_a - T_b)}{2 (1 - \mu)} \left[\frac{c + 1}{2 (c - 1)} - \frac{2c^2}{c^2 - 1} \right];$$

for r = b

$$\sigma_\Theta = \sigma_z = \frac{\alpha E (T_a - T_b)}{2 (1 - \mu)} \left[\frac{c + 1}{2 (c - 1)} - \frac{2}{c^2 - 1} \right].$$

The natural logarithm may be calculated from the formula

$$\ln c = 2 \frac{c - 1}{c + 1} + \frac{2}{3} \left(\frac{c - 1}{c + 1} \right)^3.$$

Here, if c is not greater than 2, the first term alone may be used (error not greater than 5%).

If $T_a > T_b$, the signs of the stresses will be as follows:

σ_Θ, σ_z — compression for the inner regions, and tension for the outer regions,

σ_r — compression for the whole thickness of the wall, for r = a and r = b, $\sigma_r = 0$, in accordance with the boundary conditions.

The case of the thin-walled tube is discussed in paragraph 45.

Displacements. Radial displacement

$$u = \frac{1 + \mu}{1 - \mu} \frac{\alpha T}{2} \frac{r}{\ln c} \left[\left(\frac{1 - 3\mu}{1 + \mu} + \frac{a^2}{r^2} \right) \left(\frac{1}{2} - \frac{\ln c}{c^2 - 1} \right) + \frac{1}{2} \left(1 - \frac{a^2}{r^2} \right) + \ln \frac{b}{r} - \frac{a^2}{r^2} \ln c \right].$$

The displacements on the tube surfaces will be:

for r = a

$$u = \alpha T a \left(\frac{1}{2 \ln c} - \frac{1}{c^2 - 1} \right);$$

for r = b

$$u = \alpha T b \left(\frac{1}{2 \ln c} - \frac{1}{c^2 - 1} \right).$$

50. Shape of Body [5]

Long thick-walled cylinder (Fig. 46).

Temperature Effect. The temperature distribution is unsymmetric with respect to the axis, but uniform along the axis of the cylinder. The temperature may be represented as the series

$$T = \sum_{n=1}^{\infty} [(a_n r^n + b_n r^{-n}) \cos n\Theta + (c_n r^n + d_n r^{-n}) \sin n\Theta].$$

This temperature distribution law is found if the temperature on the inner and outer surfaces of the tube is an arbitrary function of the angle Θ.

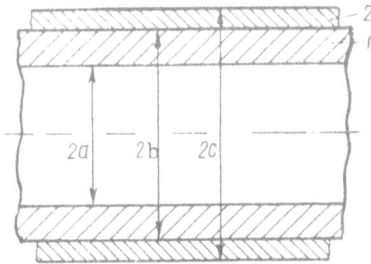

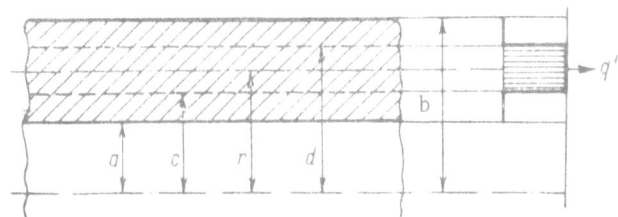

Fig. 47. Two-layer compound tube: (1) inner layer; (2) outer layer.

Fig. 48. Section along axis of tube. Heat generated in the annular volume shown by the dotted lines.

Stresses

$$\sigma_r = -\frac{\alpha E}{2(1-\mu)} \frac{r}{a^2 + b^2} \left(1 - \frac{a^2}{r^2}\right) \left(\frac{b^2}{r^2} - 1\right) (b_1 \cos\Theta + d_1 \sin\Theta);$$

$$\sigma_\Theta = \frac{\alpha E}{2(1-\mu)} \frac{r}{a^2 + b^2} \left(3 - \frac{a^2 + b^2}{r^2} - \frac{a^2 b^2}{r^4}\right) (b_1 \cos\Theta + d_1 \sin\Theta);$$

$$\tau_{r\Theta} = -\frac{\alpha E}{2(1-\mu)} \frac{r}{a^2 + b^2} \left(1 - \frac{a^2}{r^2}\right) \left(\frac{b^2}{r^2} - 1\right) (b_1 \sin\Theta - d_1 \cos\Theta);$$

$$\sigma_z = \alpha E \left[\frac{\mu}{1-\mu} \frac{r}{a^2 + b^2} \left(2 - \frac{a^2 + b^2}{r^2}\right) (b_1 \cos\Theta + d_1 \sin\Theta) - T \right].$$

Only the axial stress σ_z depends on the complete distribution law of the temperature T, while the remaining stresses depend only on the temperature, as given by the law

$$T = \frac{b_1}{r} \cos\Theta + \frac{d_1}{r} \sin\Theta.$$

The resultant axial force, calculated from the formula for σ_z, is equal to zero.

For a disc with a hole in the center (plane stressed state), $\sigma_z = 0$, while the stresses σ_Θ, σ_r, and $\tau_{r\Theta}$ are of the same form, but without the expression $(1 - \mu)$ in the denominator.

51. Shape of Body [15]

Two-layer compound tube with open ends (Fig. 47). The subscripts 1 and 2 refer to the inner and outer layers of the tube, respectively.

Temperature Effect. Transient axially symmetric temperature distribution:

$$T = T(r, t).$$

Stresses. In the inner layer 1 ($a \leq r \leq b$):

$$\sigma_\Theta = \frac{\alpha_1 E_1}{1-\mu_1} \left[\left(1 + \frac{a^2}{r^2}\right) \frac{p}{g} \int_a^b T(r, t)\, r\, dr + \frac{1}{r^2} \int_a^r T(r, t)\, r\, dr \right.$$

$$\left. + \frac{\alpha_2}{\alpha_1} \frac{1-\mu_1}{g} b^2 \left(1 + \frac{a^2}{r^2}\right) T(b, t) - T(r, t) \right];$$

$$\sigma_r = \frac{\alpha_1 E_1}{1-\mu_1} \left[\left(1 - \frac{a^2}{r^2}\right) \frac{p}{g} \int_a^b T(r, t)\, r\, dr - \frac{1}{r^2} \int_a^r T(r, t)\, r\, dr \right.$$

$$+ \frac{a_2}{a_1} \frac{1 - \mu_1}{g} b^2 \left(1 - \frac{a^2}{r^2} \right) T(b, t) \Bigg];$$

$$\sigma_z = \frac{a_1 E_1}{1 - \mu_1} \left[\frac{2}{b^2 - a^2} \int_a^b T(r, t) r \, dr - T(r, t) \right].$$

In the outer layer 2 (b ≤ r ≤ c):

$$\sigma_\theta = - \frac{a_1 E_1}{1 - \mu_1} \frac{b^2}{c^2 - b^2} \left(1 + \frac{c^2}{r^2} \right)$$

$$\times \left\{ \left[\left(1 - \frac{a^2}{b^2} \right) \frac{p}{g} - \frac{1}{b^2} \right] \int_a^b T(r, t) r \, dr + \frac{a_2}{a_1} (1 - \mu_1) \frac{b^2 - a^2}{g} T(b, t) \right\};$$

$$\sigma_r = \sigma_\theta \frac{1 - \dfrac{c^2}{r^2}}{1 + \dfrac{c^2}{r^2}},$$

where

$$p = \frac{E_1}{E_2} \left(\frac{c^2 + b^2}{c^2 - b^2} + \mu_2 \right) + \mu_1 \frac{b^2 + a^2}{b^2 - a^2} - 1;$$

$$g = \frac{E_1}{E_2} (b^2 - a^2) \left(\frac{c^2 + b^2}{c^2 - b^2} + \mu_2 \right) + b^2 (1 - \mu_1) + a^2 (1 + \mu_1).$$

52. Shape of Body [3]

Solid circular cylinder of diameter 2b.

Temperature Effect. Heat uniformly generated inside the material of the tube throughout the whole volume of the rod:

$$T = \frac{q}{4\lambda} (b^2 - r^2).$$

Stresses

$$\sigma_\theta = \frac{\alpha E q}{16 (1 - \mu) \lambda} (3r^2 - b^2);$$

$$\sigma_r = \frac{\alpha E q}{16 (1 - \mu) \lambda} (r^2 - b^2);$$

$$\sigma_z = \frac{\alpha E q}{8 (1 - \mu) \lambda} (2r^2 - b^2).$$

Displacements. In the radial direction:

$$u = \frac{\alpha q r}{16 (1 - \mu) \lambda} \left[(1 - 3\mu) b^2 + (1 + \mu)(2b^2 - r^2) \right].$$

53. Shape of Body [8]

Thick-walled tube (Fig. 48).

Temperature Effect. Heat is generated in the material of the tube in the region c ≤ r ≤ d at the constant specific power q. The heat is removed from the inner surface of the tube. The outer surface of the tube is insulated.

In general form

$$T = T_a + \frac{1}{\lambda} \int_b^a \frac{dr}{r} \int_b^r q r \, dr - \frac{1}{\lambda} \int_b^r \frac{dr}{r} \int_b^r q r \, dr.$$

The temperature drop between the inner and outer surfaces of the tube will be:

$$T_b - T_a = \frac{q}{4\lambda}(d^2 - c^2)\left(d^2 \frac{2\ln\frac{d}{c}}{d^2 - c^2} - 1 + 2\ln\frac{c}{a}\right).$$

Stresses. The maximum peripheral stress will be on the inner surface of the tube at r = a:

$$\sigma_\theta \, \text{max} = \frac{\alpha E q}{4(1-\mu)\lambda} \frac{d^2 - c^2}{b^2 - a^2}\left[\left(\frac{2d^2 \ln\frac{d}{c}}{d^2 - c^2} - 1 + 2\ln\frac{c}{a}\right)b^2 + \frac{d^2 + c^2 - 2a^2}{2}\right].$$

If heat is generated over the whole thickness of the tube (d = b, c = a), the maximum stress varies between the limits

$$\frac{2}{3}\frac{\alpha E(T_b - T_a)}{1-\mu} \leqslant \sigma_\theta \, \text{max} \leqslant \frac{\alpha E(T_b - T_a)}{1-\mu}.$$

The maximum values of the stress are for very thin-, and very thick-walled tubes.

54. Shape of Body

Thick-walled tube. Here and in the next two cases, we use the notation c = b/a, and ρ = r/a.

Temperature Effect. Heat is uniformly generated by the material in the wall of the tube at the specific power q. Cooling from inside. Insulated outside (Fig. 49).

$$T = \frac{q}{4\lambda}\left(2b^2 \ln\frac{r}{a} - r^2 + a^2\right).$$

The temperature reaches its maximum value at the outer surface of the tube where r = b.

At the same time, T_{max} at r = b is the temperature drop between the inner and outer surfaces of the tube, since it is assumed that T = 0 at the inner surface.

Stresses

$$\sigma_\theta = \frac{\alpha E q}{4(1-\mu)\lambda}\left[\frac{\rho^2 + 1}{\rho^2}\left(\frac{b^2 c^2 \ln c}{c^2 - 1} - \frac{3b^2 - a^2}{4}\right) - \frac{\rho^2 - 1}{4\rho^2}(r^2 - a^2 + 2b^2) + r^2 - a^2 - b^2 \ln\rho\right];$$

$$\sigma_r = \frac{\alpha E q}{4(1-\mu)\lambda}\left[\frac{\rho^2 - 1}{\rho^2}\left(\frac{b^2 c^2 \ln c}{c^2 - 1} - \frac{3b^2 - a^2}{4}\right) + \frac{\rho^2 - 1}{4\rho^2}(r^2 - a^2 + 2b^2) - b^2 \ln\rho\right];$$

$$\sigma_z = \frac{\alpha E q}{4(1-\mu)\lambda}\left[\frac{2b^2 c^2 \ln c}{c^2 - 1} - b^2(1 + 2\ln\rho) - \frac{b^2 - a^2}{2} + r^2 - a^2\right].$$

Figure 50 gives stress curves for c = 2.

The maximum stresses occur at the inner surface of the tube with r = a:

$$\sigma_\theta = \sigma_z = \frac{\alpha E q}{4(1-\mu)\lambda}\left(\frac{2b^2 c^2 \ln c}{c^2 - 1} - \frac{b^2 + a^2}{2}\right).$$

55. Shape of Body [3]

Thick-walled tube, c = b/a.

Temperature Effect. Heat uniformly generated by the material in the wall of the tube at specific power q. Cooling external, insulated inside.

$$T = \frac{q}{4\lambda}\left(b^2 - r^2 - 2a^2 \ln\frac{b}{r}\right).$$

The temperature reaches its maximum value at the inner surface of the tube with r = a (Fig. 51).

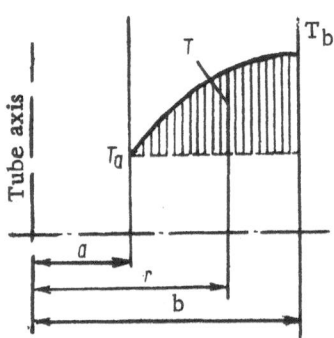

Fig. 49. Radial variation of tube temperature.

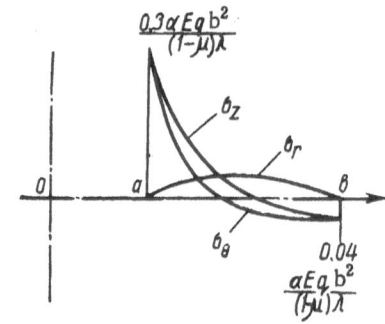

Fig. 50. Stress change along radius of tube, for the case of volume heat generation (heat removed from inside, insulated outside).

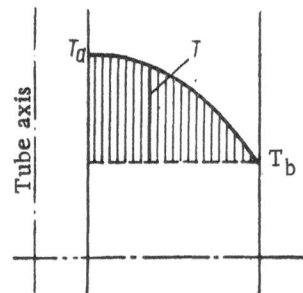

Fig. 51. Radial temperature change in tube (insulated inside, heat removed outside).

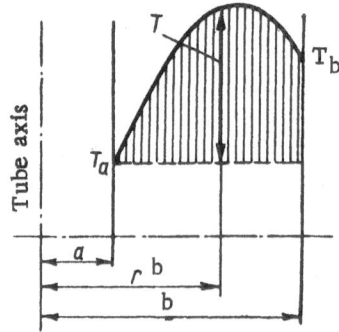

Fig. 52. Radial temperature change in tube (heat removed inside and outside).

At the same time, T_{max} at r = 0 is the temperature drop between the inner and outer surfaces of the tube, since it is assumed that T = 0 on the outer surface.

<u>Stresses</u>

$$\sigma_\theta = \frac{\imath Eq}{4(1-\mu)\lambda}\left[a^2\ln\frac{b}{r} + \frac{1+\frac{b^2}{r^2}}{c^2-1}a^2\ln c - \frac{r^2+a^2}{4}\left(1+\frac{b^2}{r^2}\right) + r^2 - a^2\right];$$

$$\sigma_r = \frac{\imath Eq}{4(1-\mu)\lambda}\left[a^2\ln\frac{b}{r} - \frac{\frac{b^2}{r^2}-1}{c^2-1}a^2\ln c - \frac{r^2-a^2}{4}\left(\frac{b^2}{r^2}-1\right)\right];$$

$$\sigma_z = \frac{\imath Eq}{4(1-\mu)\lambda}\left(\frac{2a^2\ln c}{c^2-1} + 2a^2\ln\frac{b}{r} - \frac{b^2+3a^2}{2} + r^2\right).$$

The maximum peripheral stress occurs at the outer surface of the tube with r = b:

$$\sigma_\theta\,max = \sigma_z\,max = \frac{\imath Eq}{4(1-\mu)\lambda}\left(\frac{2a^2\ln c}{c^2-1} + \frac{b^2-3a^2}{2}\right).$$

56. Shape of Body

Thick-walled tube, c = b/a, ρ = r/a.

Temperature Effect. Heat is uniformly generated by the material in the walls. Cooling inside and outside. Temperature of outer surface T_b, of inner surface T_a (Fig. 52). Since heating to constant temperature produces no stresses, T_a is subtracted from the overall temperature distribution law:

$$T = \frac{q}{4\lambda} \left[\frac{b^2 - a^2}{\ln c} \ln \rho - r^2 + a^2 \right] + \frac{T_b - T_a}{\ln c} \ln \rho.$$

Stresses

$$\sigma_\Theta = \frac{\alpha E}{1 - \mu} \left\{ \frac{q}{8\lambda} \left[\left(1 + \frac{a^2}{r^2} \right) \left(\frac{B}{\ln c} - \frac{b^2 - a^2}{2} \right) + \frac{b^2 - a^2}{r^2} \frac{R}{\ln c} \right. \right.$$

$$\left. + \frac{r^2 - a^2}{2} \left(3 + \frac{a^2}{r^2} \right) - 2 \frac{b^2 - a^2}{\ln c} \ln \rho \right] + \frac{T_b - T_a}{2 \ln c} \left[\left(1 + \frac{a^2}{r^2} \right) \frac{B}{b^2 - a^2} + \frac{R}{r^2} - 2 \ln \rho \right] \right\};$$

$$\sigma_r = \frac{\alpha E}{1 - \mu} \left\{ \frac{q}{8\lambda} \left[\left(1 - \frac{a^2}{r^2} \right) \left(\frac{B}{\ln c} - \frac{b^2 - a^2}{2} \right) - \frac{b^2 - a^2}{r^2} \frac{R}{\ln c} + \frac{(r^2 - a^2)^2}{2r^2} \right] \right.$$

$$\left. + \frac{T_b - T_a}{2 \ln c} \left[\left(1 - \frac{a^2}{r^2} \right) \frac{B}{b^2 - a^2} - \frac{R}{r^2} \right] \right\};$$

$$\sigma_z = \frac{\alpha E}{1 - \mu} \left\{ \frac{q}{4\lambda} \left[\frac{B}{\ln c} - \frac{b^2 - a^2}{2} \left(1 + 2 \frac{\ln \rho}{\ln c} \right) + r^2 - a^2 \right] + \frac{T_b - T_a}{\ln c} \left(\frac{B}{b^2 - a^2} - \ln \rho \right) \right\},$$

where

$$B = b^2 \ln c - \frac{b^2 - a^2}{2};$$

$$R = r^2 \ln \rho - \frac{r^2 - a^2}{2}.$$

The maximum peripheral stress occurs either on the inner surface (r = a), or on the outer surface of the tube (r = b):

for r = a

$$\sigma_\Theta = \frac{\alpha E q}{8(1 - \mu)\lambda} \left(b^2 + a^2 - \frac{b^2 - a^2}{\ln c} \right) + \frac{\alpha E (T_b - T_a)}{2(1 - \mu)} \left(\frac{2b^2}{b^2 - a^2} - \frac{1}{\ln c} \right);$$

for r = b

$$\sigma_\Theta = \frac{\alpha E q}{8(1 - \mu)\lambda} \left(b^2 + a^2 - \frac{b^2 - a^2}{\ln c} \right) + \frac{\alpha E (T_b - T_a)}{2(1 - \mu)} \left(\frac{2a^2}{b^2 - a^2} - \frac{1}{\ln c} \right).$$

57. Shape of Body

Long thick-walled tube. The tension diagrams of the material of the tube are known at different temperatures.

Temperature Effect. Axially symmetric temperature distribution, arbitrary along radius of tube.

Stresses. To find the radial stress distribution in the tube when part of the tube passes to the plastic state, the method of successive approximations is used in the following order [16]:

1. The thickness of the tube is divided into m equal parts, and then all the quantities given below are found for the series of radii r forming the boundaries of the segments.

2. Find the temperature T from the known distribution law.

3. Calculate the elastic stresses σ_{Θ_0}, σ_{r_0}, and σ_{z_0} (zero approximation) from the formulas already given which apply to the case in question.

4. Find the mean stress

$$\sigma_0 = \frac{2}{3} \sigma_{z_0}.$$

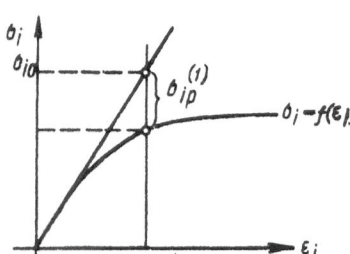

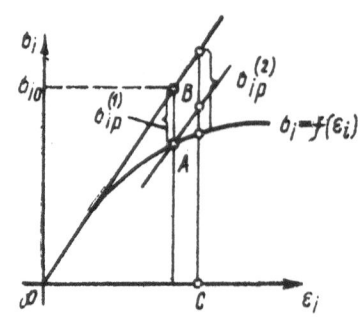

Fig. 53. Graphical determination of the magnitude of the stresses in the plastic region for a first approximation.

Fig. 54. Graphical determination of the magnitude of the stresses in the plastic region for a second approximation.

5. Find the magnitude of the stresses

$$\sigma_{i_0} = \frac{1}{\sqrt{2}} \sqrt{(\sigma_{z_0} - \sigma_{\Theta_0})^2 + (\sigma_{\Theta_0} - \sigma_{r_0})^2 + (\sigma_{r_0} - \sigma_{z_0})^2}.$$

6. From the tension diagram of the sample, $\sigma_i = f(\varepsilon_i)$ for a given temperature and given r find $\sigma_{ip}^{(1)}$ graphically (Fig. 53).

7. Find

$$\sigma_{ip}^{(1)} = \frac{\sigma_{ip}^{(i)}}{\sigma_{i_0}} (\sigma_{j_0} - \sigma_0), \text{ where } j = \Theta, z, r.$$

8. Calculate the operators $L_j^{(1)}$:

$$L_z = \frac{2D}{1 - \beta^2} - \frac{\mu A(b)}{(1-\mu)(1-\beta^2)} + \frac{\mu B(b)}{(1-\mu)(1-\beta^2)} - \frac{\mu B(r)}{1-\mu} + \frac{\mu \sigma_{rp}(r)}{1-\mu};$$

$$L_\Theta = \frac{(1-2\mu)A(b)}{2(1-\mu)(1-\beta^2)} \left(1 + \frac{a^2}{r^2}\right) + \frac{B(b)}{2(1-\mu)(1-\beta^2)} \left(1 + \frac{a^2}{r^2}\right) + \frac{(1-2\mu)A(r)}{2(1-\mu)} - \frac{B(r)}{2(1-\mu)} + \frac{\mu \sigma_{rp}(r)}{1-\mu};$$

$$L_r = -\frac{(1-2\mu)A(b)}{2(1-\mu)(1-\beta^2)} \left(1 - \frac{a^2}{r^2}\right) + \frac{B(b)}{2(1-\mu)(1-\beta^2)} \left(1 - \frac{a^2}{r^2}\right) - \frac{(1-2\mu)A(r)}{2(1-\mu)} - \frac{B(r)}{2(1-\mu)} + \sigma_{rp}(r),$$

where

$$D = \frac{1}{b^2} \int_a^b r \sigma_{zp}(r) \, dr;$$

$$\beta = \frac{a}{b};$$

$$A(r) = \frac{1}{r^2} \int_a^r r \left[\sigma_{\Theta p}(r) + \sigma_{rp}(r)\right] dr;$$

$$B(r) = \int_a^r \frac{1}{r} \left[\sigma_{\Theta p}(r) - \sigma_{rp}(r)\right] dr$$

(to reduce the amount of calculation, set $\mu = \frac{1}{2}$ for the plastic regions).

56

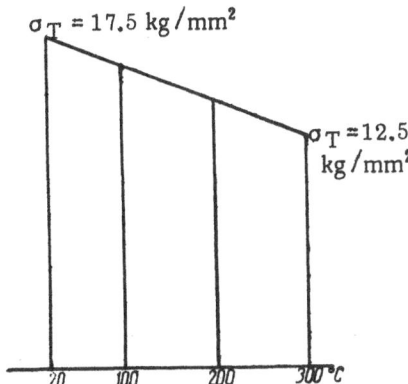

Fig. 55. Flow limit
as a function of temperature.

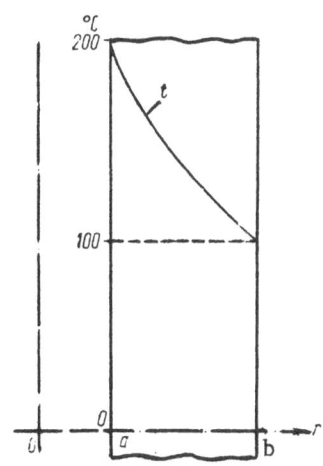

Fig. 56. Radial temperatures
in tube.

9. Find the first approximation:

$$\sigma_j^{(1)} = \sigma_{j_0} + L_j^{(1)} - \sigma_{jp}^{(1)}.$$

The solution $\sigma^{(1)}$ satisfies the equilibrium equations, the boundary conditions, and the compatibility equations for the deformations, and approximately satisfies the physical law $\sigma_j = f(\varepsilon_i)$.

The calculation can also be made without L_j, but then the solution $\sigma_j^{(1)} = \sigma_{j_0} - \sigma_{jp}^{(1)}$ will not exactly satisfy the boundary conditions.

10. Find $\sigma_i^{(1)}$ in the first approximation:

$$\sigma_i^{(1)} = \frac{1}{\sqrt{2}} \sqrt{(\sigma_z^{(1)} - \sigma_\theta^{(1)})^2 + (\sigma_\theta^{(1)} - \sigma_r^{(1)})^2 + (\sigma_r^{(1)} - \sigma_z^{(1)})^2}.$$

11. Find the error in the first approximation:

$$\Delta^{(1)} = \frac{\sigma_i^{(1)} - \sigma_i}{\sigma_i},$$

where σ_j is found graphically from the tension diagram $\sigma_i = f(\varepsilon_i)$.

If $\Delta^{(1)} = 0$, the solution is exact; if $\Delta^{(1)}$ is large, find the second approximation.

Neglecting work hardening, σ_i is the same as σ_T, and then, instead of finding the value of Δ, curves may be constructed showing the change in σ_T and $\sigma_i^{(1)}$ over the thickness of the tube wall, from which the degree of agreement between $\sigma_i^{(1)}$ and σ_T may be found visually. We should have: $\sigma_i^{(1)} \le \sigma_i = \sigma_T$.

12. Make a graphical determination of $\sigma_{ip}^{(2)}$ in the second approximation (Fig. 54).

Draw a line from point A parallel to the elastic part of the tension diagram to point B, as given by the condition BC = $\sigma_i^{(1)}$. The desired value of $\sigma_{ip}^{(2)}$ is shown in Fig. 54.

13. Find

$$\sigma^{(1)} = \frac{\sigma_z^{(1)} + \sigma_\theta^{(1)} + \sigma_r^{(1)}}{3}.$$

14. Find

$$\sigma_{jp}^{(2)} = \frac{\sigma_{ip}^{(2)}}{\sigma_i^{(1)}} (\sigma_j^{(1)} - \sigma^{(1)}).$$

57

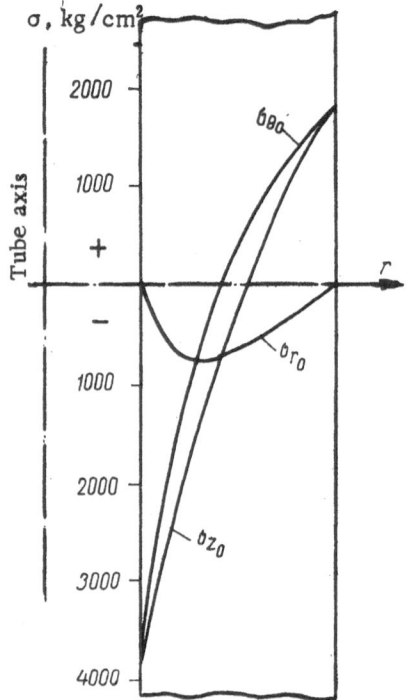

Fig. 57. Radial stress change in tube in the zero approximation (elastic solution).

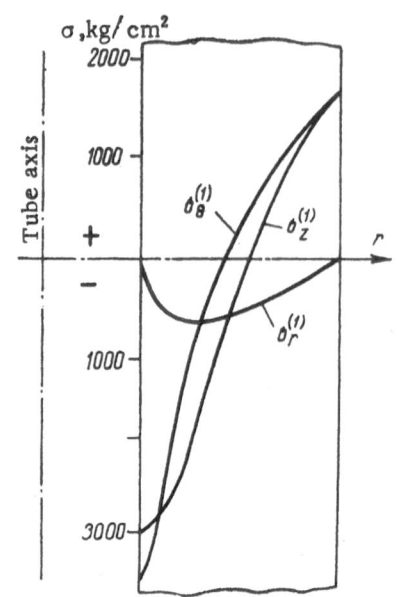

Fig. 58. Radial stress change in tube in first approximation.

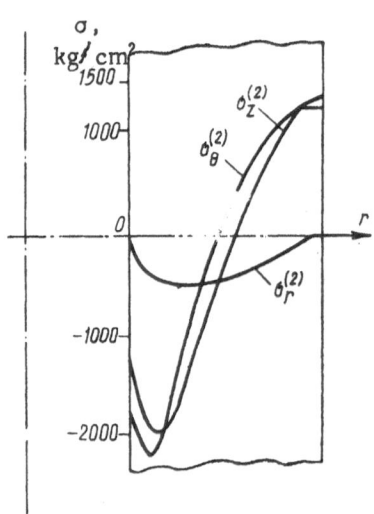

Fig. 59. Radial stress change in tube in second approximation.

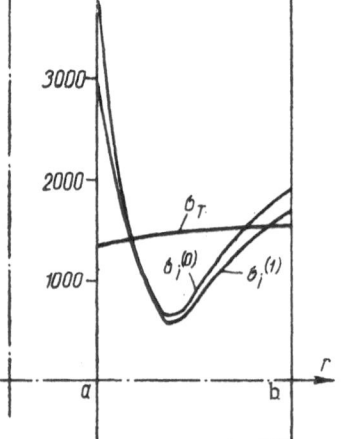

Fig. 60. Stresses (zero and first approximation) and flow limit over thickness of tube.

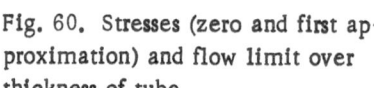

Fig. 61. Stresses (second approximation) and flow limit over thickness of tube.

15. Calculate the operators $L_j^{(2)}$ for the second approximation from the same formulas (paragraph 7).

16. Find the second approximation:

$$\sigma_j^{(2)} = \sigma_j^{(1)} + L_j^{(2)} - \sigma_{j\rho}^{(2)}.$$

17. Find

$$\sigma_i^{(2)} = \frac{1}{\sqrt{2}} \sqrt{(\sigma_z^{(2)} - \sigma_\theta^{(2)})^2 + (\sigma_\theta^{(2)} - \sigma_r^{(2)})^2 + (\sigma_r^{(2)} - \sigma_z^{(2)})^2}.$$

TABLE 6. Calculation of Thermal Stress in a Tube by Successive Approximations

r	2	2.5	3	3.5	4	4.5	5	5.5	6
$\ln r$	0.6931	0.9163	1.0986	1.2527	1.3863	1.5041	1.6094	1.7047	1.7918
T	100	80	63	49	37	26.2	16.6	7.9	0
σ_T	1390	1420	1450	1460	1470	1500	1510	1530	1540
σ_{z_0}	−3830	−2690	−1710	−914	−228	390	936	1440	1890
σ_{Θ_0}	−3830	−2100	−980	−212	377	865	1250	1590	1890
σ_{r_0}	0	−590	−735	−703	−607	−472	−315	−155	0
σ_0	−2550	−1790	−1140	−610	−152	260	624	960	1260
σ_{i_0}	3830	1870	880	624	860	1170	1430	1680	1890
$\sigma_{ip}^{(1)}$	2440	450	0	0	0	0	0	150	350
$\sigma_{z_0} - \sigma_0$	−1280	−900	−	−	−	−	−	480	630
$\sigma_{\Theta_0} - \sigma_0$	−1280	−310	−	−	−	−	−	630	630
$\sigma_{r_0} - \sigma_0$	2550	1200	−	−	−	−	−	−1115	−1260
$\sigma_{zp}^{(1)}$	−815	−217	−	−	−	−	−	43	116
$\sigma_{\Theta p}^{(1)}$	−815	−75	−	−	−	−	−	56	116
$\sigma_{rp}^{(1)}$	−1620	290	−	−	−	−	−	−100	−233
$\sigma_{\Theta r} - \sigma_{rp}$	−2435	−1015	0	0	0	0	0	156	349
$\dfrac{\sigma_{\Theta p} - \sigma_{rp}}{r}$	−1215	−418	0	0	0	0	0	28.4	58
$B(r)$	0	−410	−512	−512	−512	−512	−512	−505	−482
$r\sigma_{zp}$	−1630	−543	0	0	0	0	0	236	695
D	−	−	−	−	−	−	−	−	−10.7
$\sigma_{\Theta p} + \sigma_{rp}$	805	215	0	0	0	0	0	−44	−116
$r(\sigma_{\Theta p} + \sigma_{rp})$	1610	538	0	0	0	0	0	−242	−695
$A(r)$	0	86	75	55	42	33	27	20	8.5
L_z	294	18.4	−37	−37	−37	−37	−37	−70	−116
L_Θ	−237	−208	−161	−125	−103	−87	−76	−103	−153
L_r	1620	299	113	78	46	41	31	−82	−233
$\sigma_z^{(1)}$	−2721	−2454	−1747	−951	−265	353	899	1327	1658
$\sigma_\Theta^{(1)}$	−3252	−2233	−1141	−337	274	778	1174	1431	1621
$\sigma_r^{(1)}$	0	−581	−622	−625	−561	−431	−284	−137	0
$\sigma_i^{(1)}$	3150	1780	970	530	730	1062	1340	1520	1720
$\Delta^{(1)}$	1.26	0.25	−	−	−	−	−	−	0.12
$\sigma_{ip}^{(2)}$	4200	810	0	0	0	0	0	140	530

TABLE 6 (continued)

r	2	2.5	3	3.5	4	4.5	5	5.5	6
$\sigma_{zp}^{(2)}$	−970	−318	0	0	0	0	0	41.5	175.0
$\sigma_{\theta p}^{(2)}$	−1670	−213	0	0	0	0	0	51.7	163
$\sigma_{rp}^{(2)}$	2640	535	0	0	0	0	0	−93	−333
L_z	599	151	14	14	14	14	−14	−15	−97.5
L_θ	−155	−238	−190	−154	−125	−104	−89	−112	−196
L_r	2640	726	153	125	74	53	39	−69	−333
$\sigma_z^{(2)}$	−1152	−1985	−1733	−937	−251	367	913	1271	1385
$\sigma_\theta^{(2)}$	−1737	−2258	−1331	−492	149	674	1085	1267	1262
$\sigma_r^{(2)}$	0	−390	−469	−500	−487	−378	−245	−114	0
$\sigma_i^{(2)}$	1550	1600	1130	437	557	923	1255	1390	1330
$\Delta^{(2)}$	0.115	0.127	−	−	−	−	−	−	−

18. Find the error in the second approximation

$$\Delta^{(2)} = \frac{\sigma_i^{(2)} - \sigma_i}{\sigma_i},$$

where $\sigma_i^{(2)}$ and σ_i occur at the same value of ε_i on the tension diagram. If $\Delta^{(2)}$ is large, find the third approximation, etc., in the same way, until the error, $\Delta^{(n)}$, becomes sufficiently small.

Example. Take a thick-walled, very long tube, subjected to a thermal load consisting of a steady heat flux from inside through the wall of the tube. The surface temperature of the wall is 200°C inside, and 100°C outside.

Since uniform heating to 100°C does not produce any macrostresses, the calculation is made from the condition that the temperature drop between the inner and outer surfaces of the tube walls is 200 − 100 = 100°C.

The inner radius is a = 2 cm, and the outer radius is b = 6 cm.

The material of the tube is uranium, $E = 2 \cdot 10^6$ kg/cm², $\alpha = 22 \cdot 10^{-6}$ deg⁻¹ at 300°C, $\sigma_T = 12.5$ kg per mm² at 20°C, $\sigma_T = 17.5$ kg/mm², and $\mu = 0.23$. We assume no work hardening (ideal plasticity diagram). Accordingly, in the plastic region, $\sigma_i = \sigma_T$.

The relation assumed to exist between the flow limit and the temperature is shown in Fig. 55. The radial temperature change in the tube is shown in Fig. 56.

The results of the calculation are given in Table 6. The dimensions are everywhere cm, deg, and kg. The integrals in the expressions for D, A, and B are evaluated from the formula

$$\int_a^r f\,dr = \frac{\Delta r}{2}\,[f_0 + 2f_1 + 2f_2 + f_3].$$

Figures 57 to 59 show curves for the zero, first, and second approximations. Stress curves for the zero and first approximations are shown in Fig. 60, and for the second approximation in Fig. 61. The same curves show how the flow limit varies over the thickness of the tube wall. In the plastic zones, σ_i approaches σ_T in each approximation.

LITERATURE CITED

1. Davidenkov, N. N., "Residual stresses, in: X rays as Applied to the Study of Materials, Leningrad, ONTI, 1936.
2. Boas and Honeycombe, Proc. Roy. Soc. (London) 186:57 (1946); Inst. of Metals 73:433 (1946).
3. Freudenthal, A. M., Nuclear Engineering, Bonilla, 1957, Ch. II, Thermal Stress Analysis and Mechanical Design.
4. Schmid, F., and Linther, K., "Metallkundliche Probleme beim Bau von Reaktoren," Z. Metallk. 47(4):276 (1956).
5. Melan, É. and Parkus, G., Thermoelastic Stresses Produced by Steady State Temperature Fields, Moscow, Fizmatgiz, 1958.
6. Goodier, I. N., Thermal Stress and Deformations, J. Appl. Mech. 24(3):380 (1957).
7. Sharma, B., "Thermal stress in infinite elastic disks," J. Appl. Mech. 23(4):527 (1956).
8. Thompson, A. S., and Rodgers, O. E., Thermal Power from Nuclear Reactors, New York, 1956.
9. Fridman, Ya. B., and Morozov, E. M., "The approximate calculation of stress concentration in compound samples," Scientific Reports of the Higher School, Machine and Instrument Construction (4), 1958.
10. Huth, J. Appl. Phys. 23:1234 (1952).
11. Gatewood, B. E., Thermal Stresses, IL, 1959.
12. Boley, B. A., "The determination of temperature stress and deflections in two-dimensional thermoelastic problems," J. Aeronaut. Sci. 23(1):67 (1956).
13. Smith, Nuclear Sci. Eng. 2(3):152 (1957).
14. Weiner, I., "An elastoplastic thermal stress analysis of a free plate," J. Appl. Mech. 23(3):395 (1956).
15. Neubauer, Temperatur und Spannungsverteilung in Zylindrischen Körper, Berlin, 1958.
16. Shorr, B. V., Calculation of Nonuniformly Heated Cylinders in the Elastoplastic Range, Izd. AN SSSR, No. 6, 1960.

THERMAL FATIGUE AND THERMAL SHOCK

N. D. Sobolev and V. I. Egorov

STRENGTH OF MATERIALS UNDER VARYING TEMPERATURES

Introduction

Problems in the strength of structural materials in a variable temperature field are attracting more and more attention from engineers, as well as from investigators working in the most diverse branches of technology. A number of reviews [1-3] have appeared recently and are used in the present paper.

Stresses and deformations from temperature effects are to a considerable extent responsible for the failure of molds, rolls, plungers, and dies, as well as petroleum refining and railroad equipment [4-13].

Strength problems under changing temperatures are especially important in some of the newer branches of engineering: in modern aviation technology, which has produced airplanes that fly at supersonic speed, in which kinetic heating occurs, in rocket technology, since startup of rockets requires generating a large amount of heat per unit time, and subsequent flight occurs at very high velocities, and in reactor construction, where the large heat fluxes in the fuel elements of nuclear reactors produce temperature gradients which cause large temperature stresses to occur.

Strength problems under varying temperatures are no less important in other engineering operations. An example is provided by the treatment of metals and ceramic materials at elevated temperatures.

A special feature of the behavior of materials in a variable temperature field is that thermal stresses and deformations (often of considerable magnitude) occur under conditions where the mechanical, physical, and chemical properties of the material can change substantially. The problem is further complicated by the fact that, as a rule, mechanical stresses from external loads are acting along with the thermal stresses.

The importance of making a thorough-going study of the mechanical behavior of materials under changing temperatures is growing steadily, and better methods of analytical and experimental study must be developed.

The problem takes on particular importance in view of the tendency toward increasing the temperatures and heat fluxes, as well as increasing the dimensions of the parts used in nuclear energy installations.

Conditions Under Which Thermal Deformations and Stresses Occur

The result of a changing temperature field is to produce thermal deformations and stresses, which arise from the static indeterminacy of the system under the following conditions:

1. Nonuniform heating or cooling of a body when it is not connected to any other bodies;

2. Uniform heating or cooling of a body which has external mechanical restraints applied to it;

3. Nonuniform heating or cooling of a body with external mechanical restraints.

While in the first and third cases the deformation occurs as the result of a temperature gradient in the body, in the second case the temperature distribution in the body is uniform at any instant of time.

Effect of the Nature of Thermal Stresses and Deformations on the Mechanical Behavior of Materials

It is convenient to distinguish the following possible forms of thermal loading.

1. A gradual, but unrepeated change in temperature which produces a steady state temperature field and static stresses.

The result of this may be:

a. Constant elastic deformations and stresses;

b. Constant elastoplastic stresses and deformations;

c. At relatively high stresses of long duration, creep may occur and, as a result, relaxation of the stresses. This type of creep and relaxation may be given the name of thermal creep or relaxation;

d. Failure of low plasticity materials at relatively high stresses.

2. Cyclic temperature change, which produces a variable temperature field and variable cyclic stresses.

This can produce:

a. Cyclically varying elastic stresses and deformations below the fatigue limit;

b. Cyclic stresses and deformations leading to fatigue failure. This type of fatigue failure is usually known as thermal fatigue;

c. Stresses which, under definite conditions, produce creep and, as a result, considerable change in shape, and finally fatigue failure if they act long enough. This phenomenon has entered the literature under the name of cyclic thermal treatment (CTT).

3. Large but unrepeated temperature change which produces a temperature field with a large temperature gradient and large dynamic stresses.

The result of this may be:

a. Large stresses, and correspondingly large plastic deformations in highly plastic materials;

b. Large stresses leading to brittle failure as a result of the dynamic character of the loading. This type of temperature effect has received the name of thermal shock.

In contrast with the above classification, the paper by Yu. F. Balandin [3] makes no clear distinction between thermal fatigue and CTT. Although these two phenomena have the same cause, they must be considered separately, since the processes of failure (thermal fatigue) and of deformation (change in shape in CTT) follow different laws.

It was shown in the paper by N. M. Sklyarov et al. [13] that when there is a nonuniform steady state field produced by gradual temperature change, the danger of loss of strength is, for practical purposes, determined solely by the mechanical loading, since almost complete relaxation of the thermal stresses occurs during prolonged operation.

On the other hand, the danger of failure from cyclic thermal stresses (thermal fatigue) and in sudden temperature change (thermal shock) is very great and, in the present paper, we are concentrating our attention on these problems which are extremely important from an engineering point of view. Change in shape in CTT is not discussed here.

The paper includes the following sections:

1. Strength under cyclic thermal stresses;

2. Strength under simultaneous action of cyclic thermal stresses and static loading;

3. Strength under thermal shock.

STRENGTH UNDER CYCLIC THERMAL STRESSES (THERMAL FATIGUE)

Thermal fatigue is the result of cyclic thermal stresses and deformations.

Consider a number of examples. In a pipe expansion joint with clamped ends, cyclic change in the temperature flux from the medium produces alternating deformations and stresses. Periodic temperature change in the ambient medium produces a deformation in the bimetallic element of a thermostat. Uniform change in the temperature of a polycrystal produces thermostructural stresses (stresses of the second kind) as a result of anisotropy in the thermal expansion of the crystals, or as a result of differences in properties of the structural constituents. The need to take account of these stresses is discussed below.

When there is a change in operating conditions, the housing of a nuclear reactor can undergo variable thermal stresses. A similar picture is observed in the fuel element cladding, in the heat exchanger piping, etc. Large cyclic thermal stresses can occur in the various parts of mobile reactors during startup and shutdown.

While in the first three examples the deformation at any instant of time is occurring practically with no temperature gradient in the body, in the rest of the examples the temperature is nonuniformly distributed.

The mechanism of thermal fatigue is similar to the mechanism of fatigue under mechanical action, since in both cases failure is caused by repeated stresses and deformations. However, if we compare the conditions under which thermal and mechanical fatigue occur, we see that complete correspondence does not exist between the two types of failure.

Relation Between Stresses and Deformations in Thermal Fatigue

The deformation process in an elementary volume of a material in thermal fatigue occurs from stresses that are cyclically varying in magnitude and sometimes in time, under cyclically varying temperature conditions. This affects the course of the deformation, and requires special consideration.

In order to discover the mechanical nature of the process, consider the deformation diagram for the uniaxial stressed state produced in a sample with rigidly clamped ends (Fig. 1) [1,14]. The sample is cyclically heated and cooled, which produces periodic tension and compression along with repeated thermal stresses. Under these conditions, the length of the sample remains constant, and so the mechanical and temperature deformations are equal in absolute value. If the stresses do not exceed the elastic limit, the deformation occurs in the elastic range, as shown in Fig. 2.

Consider the simplest case of the behavior of an ideally plastic material without work hardening. We assume that at the initial instant of time the sample is free of stresses and heated to the maximum temperature T_{max}.

On subsequent cooling, since the ends of the sample are clamped, tensile stresses occur with corresponding mechanical deformations, which are equal to the temperature deformations. The temperature deformation is equal to $\alpha \Delta T$, where α is the coefficient of linear expansion, and ΔT is the temperature drop.

It may be seen from Fig. 1 that if the temperature deformation is less than the deformation ε_T corresponding to the point A, there will be a cyclic change in the stresses in the elastic region. However, at higher values of the temperature deformation, plastic deformation will occur.

We shall assume that the temperature deformation is greater than ε_T, but less than $2\varepsilon_T$. In this case, plastic deformation occurs at the temperature T_{min}, determined by the segment AB. Increasing the temperature further to T_{max} removes the tensile stresses (line BC) and produces compressive stresses (line C'C). With subsequent cyclic change in temperature between the limits $T_{max} - T_{min}$, the change in stresses and deformation will correspond with the line CB, i.e., the process will take place over an unsymmetric cycle in the elastic range. The residual deformation AB resulting from the first cycle produces cold working in the material, and from then on, the process will take place without any plastic displacements.

If the temperature deformation is equal to $2\varepsilon_T$, the instant at which the temperature is lowered to T_{min} corresponds with the point D in Fig. 1. Increasing the temperature to T_{max} removes the tensile stresses and produces compressive stresses equal to σ_T (it is assumed that the flow limits are the same in tension and compression,

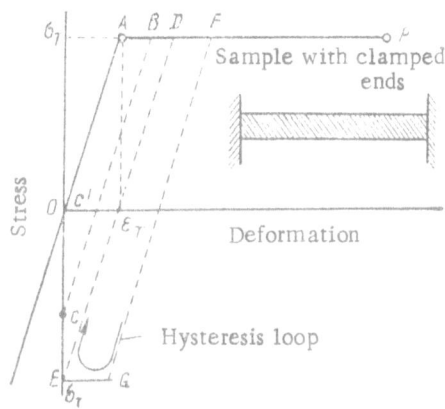

Fig. 1. Deformation diagram of a sample with clamped ends under cyclic heating and cooling [1].

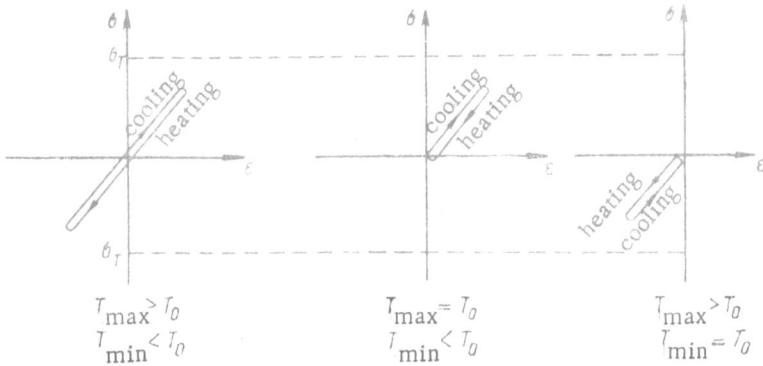

Fig. 2. Deformation diagrams of clamped sample under cyclic heating and cooling in the elastic range, with initial sample temperature T_0.

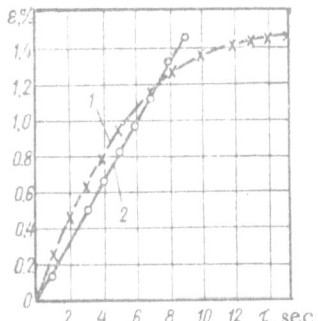

Fig. 3. Elastoplastic deformation curves [15]: (1) cooling (from 750 to 100°C); (2) heating (from 100-750°C).

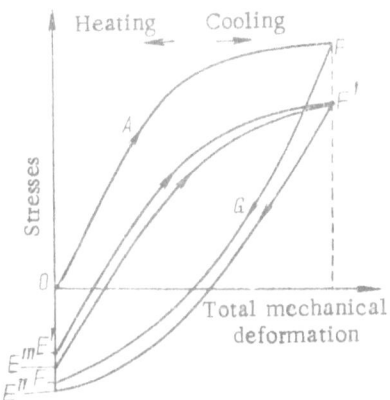

Fig. 4. Schematic deformation diagram for several temperature cycles.

and are independent of temperature). Further change in temperature from T_{max} to T_{min} will produce a change in stresses and deformations over a symmetric cycle (the line DE), also without plastic displacements.

If the temperature deformation is greater than $2\varepsilon_T$, the temperature T_{min} corresponds with the point F in Fig. 1. As the temperature is raised to T_{max}, the course followed by the deformation is given by the broken line FGE, and the subsequent cyclic temperature change $T_{max} - T_{min}$ produces the hysteresis loop EDFG, i.e., it produces plastic deformation of alternating sign, where the amplitude corresponds with the segment DF. This deformation produces irreversible changes which lead to cumulative damage that can finally produce failure.

We shall now consider the deformation of actual materials which become work hardened during plastic deformation.

Note that the relation between the stresses and deformations is continuously changing as a result of continuous change in the temperature conditions, and so the usual deformation curves cannot be used to make an exact analysis of the deformation process.

Recently, a method has been worked out for constructing diagrams of the deformation that occurs under changing temperatures [15]. The first such diagrams, taken for ÉI437B alloy, show that the deformation occurs differently in tension and compression (Fig. 3).

Account must be taken of the Bauschinger effect in regular plastic deformation, where the resistance to plastic deformation in one direction (for example, in tension) is lowered as a result of the preceding plastic deformation in the opposite direction.

Finally, the effect of stress relaxation must be kept in mind. The relaxation is greatest at the highest temperatures in the cycle.

The deformation process in an actual material is shown schematically in Fig. 4. It is assumed as before that the rigidly clamped sample was initially stress-free and heated to the temperature T_{max}. On cooling to the temperature T_{min}, the relation between stress and deformation is given by the curve OAF. If the sample is kept at T_{min}, there is relatively little or no stress relaxation. The deformation on subsequently heating the sample to T_{max} is shown by the curve FGE, which differs from the curve OAF because of the change in sign of the stresses under different temperature conditions, and because of the Bauschinger effect. During the time the sample is held at the temperature T_{max} (point E), it is possible to have stress relaxation, which shifts the point E to E' in Fig. 4. With further change in temperature, the deformation follows the curve EF'E", E"F', etc. After a definite number of cycles a more or less stable hysteresis loop is set up which gives the energy expended per cycle in building up damage.

Note in this connection that the Bauschinger effect is particularly large in the initial stage of cyclic plastic deformation, and shows up as a large reduction in $\sigma_{0.2}$ and σ_{pc}, and a small reduction in the modulus of elasticity E. It must also be pointed out that the Bauschinger effect persists for a long time in metals at room temperature.

Heating restores the initial plastic properties of the sample, as is observed at elevated temperatures in the cycle. However, restoration of the macromechanical properties by aging after each loading usually does not increase the number of cycles before failure, i.e., it does not remove the microdamage.

Quantitative Thermal Fatigue Calculations

Thermal fatigue failure is a long, drawn-out process. Accordingly, the data on the behavior of plastic materials are given as the number of cycles before failure as a function of some constant of the cycle.

The most important constants of the cycle are the following:

1. The change in elastoplastic deformation per cycle;

2. The stress change per cycle; and

3. The energy expended per cycle, as given by the area of the hysteresis loop.

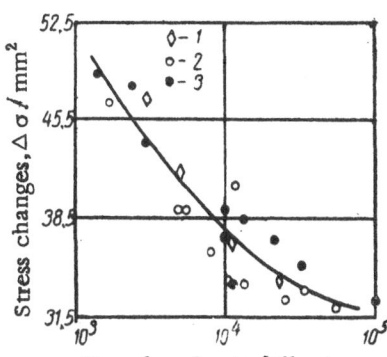

Fig. 5. Number of cycles to failure as a function of stress change (measured on the basis of 1000 cycles) in various tests [23]: (1) complete clamping − T_{max} − compression; (2) complete clamping − T_{max} − tension; (3) partial clamping − T_{max} − tension.

If the deformation occurs in the elastic range (see Fig. 2), it is a good idea to express the cycle in terms of the stress, since it is easier in this case to make a comparison with the resistance to mechanical fatigue, which is usually expressed in terms of the stresses. Since, for the majority of materials, there is no data on the fatigue limits under alternating temperatures, the following approximate estimate is to be recommended:

$$|\sigma|_{max} \leqslant \sigma_r(T_m),\qquad(1)$$

where $\sigma_r(T_m)$ is the fatigue limit at the mean temperature of the cycle.

A more conservative estimate of the strength may be made from the condition

$$|\sigma|_{max} \leqslant \sigma_r(T_{max}),\qquad(2)$$

where $\sigma_r(T_{max})$ is the fatigue limit at T_{max}.

Coffin [14] has made use of the amount of plastic deformation per cycle as a constant which gives a unique definition of the thermal fatigue under any given temperatures and conditions.

Note that this is a rather complicated characteristic. Part of the plastic deformation in the cycle occurs at elevated temperature, while the other part occurs at reduced temperature. Further, part is time dependent while the other part is not.

Accordingly, part of the plastic deformation may occur at the slip planes inside the grains, while the rest occurs at the grain boundaries.

Further, this characteristic does not by itself reflect the effect of the various factors which produce effects that show up especially at elevated temperature. Hence, it may be seen that, in addition to the amount of plastic deformation per cycle, it is necessary to know the temperature level at which the thermal fatigue is taking place.

There is a certain amount of data (Fig. 5) to show that the stresses may be taken as the criterion of strength. However, there is still not enough quantitative experimental data to make a final choice of this criterion.

Taking the energy expended per cycle as a characteristic quantity is of definite interest, since the energy gives an overall picture of the force and deformation. Since there are no experimental data at the present time, it is still difficult to say whether or not it is a good idea to make use of this quantity.

The amount of plastic deformation per cycle [14, 16] is often used as a characteristic constant of the thermal fatigue cycle.

Making an experimental measurement of the plastic deformation presents no difficulties as long as the modulus of elasticity E is independent of temperature, and there is no creep and, accordingly, no stress relaxation (Fig. 6):

$$\varepsilon_p = \varepsilon_T - \varepsilon_e = \alpha\Delta T - \frac{\Delta\sigma}{E},\qquad(3)$$

where $\Delta\sigma$ is the stress change, E is the modulus of elasticity, ε_T is the temperature deformation, and ε_e is the elastic part of the deformation.

If, however, creep is observed to occur at the time the sample is being held at T_{max}, the plastic deformation is not directly related to the total stress level. During the tension period, the plastic deformation is equal to

$$\varepsilon_{t,p} = \alpha\Delta T - \frac{\Delta\sigma_1}{E},\qquad(4)$$

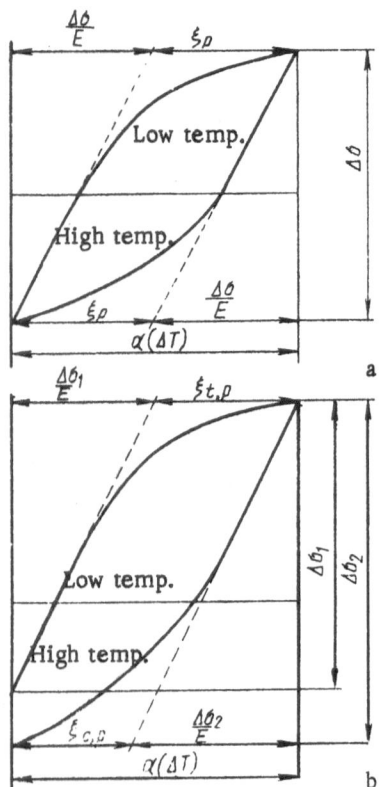

Fig. 6. Deformation diagram without (a) and with (b) stress relaxation.

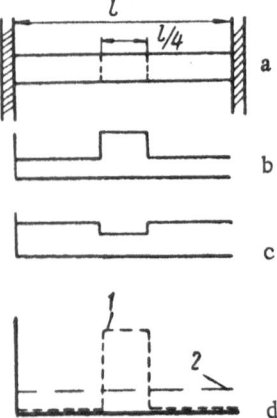

Fig. 7. Localization of plastic deformation resulting from nonuniform temperature distribution: (a) sample with ends clamped; (b) temperature distribution; (c) distribution of flow limits; (d) distribution of deformation; (1) localization of deformation; (2) mean deformation.

where $\Delta\sigma_1$ is the stress change in the cooling period.

During heating, the stress drop increases, which reduces the plastic deformation and compression, $\varepsilon'_{c,p}$.

However, the total plastic deformation includes the creep deformation $\varepsilon''_{c,p}$. Accordingly

$$\varepsilon_{c,p} = \varepsilon'_{c,p} + \varepsilon''_{c,p}$$
$$= \alpha\Delta T - \frac{\Delta\sigma_2}{E} + \frac{\Delta\sigma_2 - \Delta\sigma_1}{E} = \alpha\Delta T - \frac{\Delta\sigma_1}{E}. \tag{5}$$

Thus, the plastic deformations in compression and tension are the same, but they are not related to the total stress change, $\Delta\sigma_2$.

Usually, the amount of plastic deformation is found from the stress change. And, if the value of $\Delta\sigma_2$ is used in the calculation, errors are possible when creep is present.

Further, attention must also be given to the fact that all experiments on thermal fatigue have actually been made with an appreciable temperature gradient along the length of the sample as a result of the difference in heat transfer between the middle and the ends of the sample. If this is not taken into consideration, it is possible to make large errors in determining the actual plastic deformation.

Assume that in a sample of length l (Fig. 7), the temperature in the middle part equal to $l/4$ is higher than in the rest of the sample, and accordingly the flow limit will have its highest value in the middle. Then the plastic deformation is localized primarily in this part of the sample. In the limiting case, where the temperature of the cold part is only slightly lower than that of the hot part, but enough lower for the deformations to be elastic in the cold part, the plastic deformation in the hot part of the sample will actually be a factor of four greater than what would occur if the sample were uniformly heated. The deformation is also greater than what would occur if the end cross sections of the hot part were clamped, since the thermal deformation of the sample for its whole length l occurs in the plastic deformation of the segment $l/4$. This shows once more that equality between mechanical and thermal deformation is not an extreme case, as will be pointed out below.

In actual fact, work hardening and elastic deformation cause the situation to be less clear cut. In any case, localization of the deformation must be kept in mind when trying to account for the large difference in the number of cycles to failure between thermal fatigue and fatigue at constant temperature.

In steady-state cyclic plastic deformation, the most important thing is to set up a relation between the amount of plastic deformation and the number of cycles to failure.

We shall discuss this question primarily for cyclic plastic deformation at constant temperature.

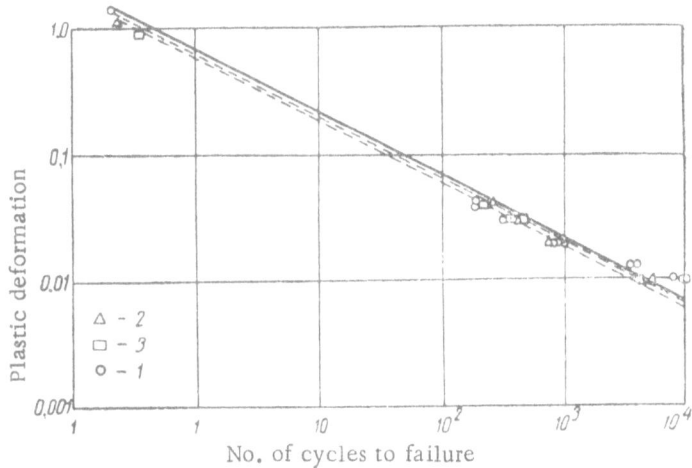

Fig. 8. Results of thermal fatigue tests as a function of amount of plastic deformation [17]: (1) 18-8 stainless steel; (2) steel containing 13% Cr; (3) chrome-molybdenum steel.

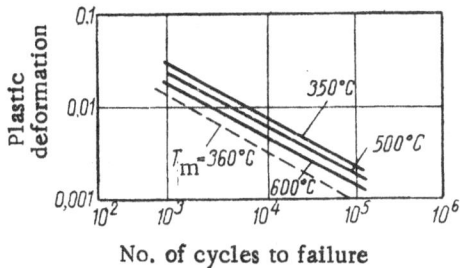

Fig. 9. Results of thermal fatigue tests under cyclically changing and constant temperatures [19]. Solid lines are for constant temperature; the dotted line is for cyclically changing temperature.

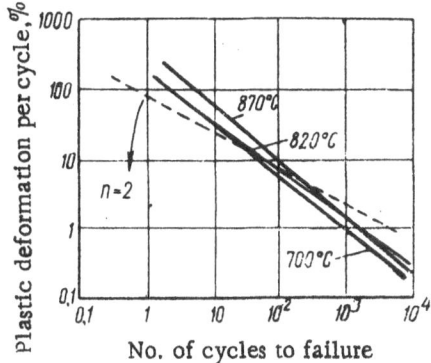

Fig. 10. Results of tests on Inconel 550 for deformation with alternating sign [20].

Since failure occurs as a result of exhausting the plasticity of the material, it seems at first glance that it might be a good idea to use the total plastic deformation, regardless of sign, as a criterion for the danger of failure.

However, this approach does not work. Thus, a material which had 50% plasticity in tension gave a total deformation of 10,000% under deformation alternating in sign.

It should, however, be emphasized that the total plastic deformation in cyclic deformation, although not equal to the plasticity under single loading, does depend on this characteristic constant. The total deformation also depends on the amount of deformation per cycle.

The relation between the number of cycles to failure and the amount of plastic deformation per cycle has been found by Manson [1] from the results of Liu's work, done on aluminum at room temperature:

$$N = \frac{K}{\varepsilon_p^n} \, . \tag{6}$$

It was found that $n \approx 2$. The value of n is also in good agreement with the results of tests on type 347 stainless steel under thermal fatigue [14].

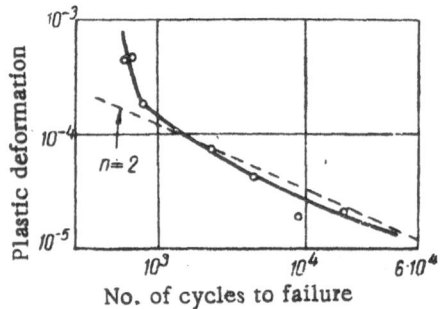

Fig. 11. Change in number of cycles to failure of S-816 alloy with change in mean plastic deformation per half cycle [21].

Coffin [17], using the results of various investigators for fatigue at room temperature, came to the conclusion that it is possible to use Eq. (6). Since, in the usual fatigue tests, the number of cycles to failure changes depending on the magnitude of the stresses, the author had to find the corresponding plastic deformations by an indirect method (Fig. 8). Similar relations have also been found for various aluminum alloys [18]. It should be noted that the results of one-time tensile tests, where $N = \frac{1}{4}$, give a good fit to the general curve in Fig. 8. This shows that it is in some cases possible to calculate the coefficient K without making fatigue tests, from the plasticity and tension with $N = \frac{1}{4}$.

The results of fatigue tests at elevated temperatures of 350, 500, and 600°C for 347 stainless steel [19] give a good fit to straight lines (Fig. 9) with $n = 2$. However, the points with $N = \frac{1}{4}$ in this case do not fall on the lines, which eliminates the possibility of calculating the coefficient K from the results of tensile tests. Obviously, the lack of fit comes from a number of effects occurring under prolonged exposure to elevated temperature.

Kennedy [20] found (Fig. 10) that in Eq. (6) for Inconel at 705, 815, and 870°C, n is equal to 1.2-1.5, which possibly comes from creep.

The relation between the number of cycles to failure and the amount of plastic deformation in thermal fatigue requires special consideration, since the deformation occurs under varying temperature.

In Fig. 9, the dotted line gives the results of tests on 347 stainless steel where the temperature changed from 200 to 500°C, giving a mean temperature of 350°C. It may be seen that for the same amount of deformation, the number of cycles to failure is less in this case than the number of cycles to failure at a constant temperature of 350 or even 600°C. This is possibly due to the fact that the deformation is localized in the sample in the thermal fatigue test, as has already been pointed out, as well as to having more extensive development of physicochemical and structural processes under variable temperature than is the case at constant temperature. The thermal fatigue endurance is a factor of 2.5 less than when the tests were made at a constant temperature of 350°C. It is possible that if the calculation were made from the actual deformations, the dotted line would be closer to the line for 350°C. However, in view of the fact that in thermal fatigue the material is at an elevated temperature (above 350°C) for a part of the cycle, where the various weakening effects are enhanced, it is to be expected that the dotted line will fall somewhat lower.

Clauss and Freeman [21] have made tests on cobalt base S-816 alloy, after aging, where T_{min} remained constant but T_{max} varied, so that the mean temperature and amount of plastic deformation changed. Figure 11 shows that for relatively small values of the deformations, and hence of the temperatures T_{max}, the points fall approximately on a line corresponding with Eq. (6).

The Nature of the Stressed State, and the Choice of a Criterion of Strength in Thermal Fatigue

The most diverse types of stressed states can occur, depending on the conditions under which the structural elements are operating.

We shall discuss in detail a number of examples of stressed states under cyclic temperature change.

In the U-shaped pipe expansion joint (Fig. 12a), uniaxially stressed states of tension and compression occur alternately under cyclic heating and cooling.

The analysis of [22] shows that an equiaxial plane stressed state occurs in the most stressed states at the surface of a thick-walled cylinder with a radial temperature drop. The stresses change sign under cyclic heating and cooling, and the plane tension stressed state is replaced by a plane compression stressed state. This is the picture which occurs in the piping of heat exchangers (Fig. 12b).

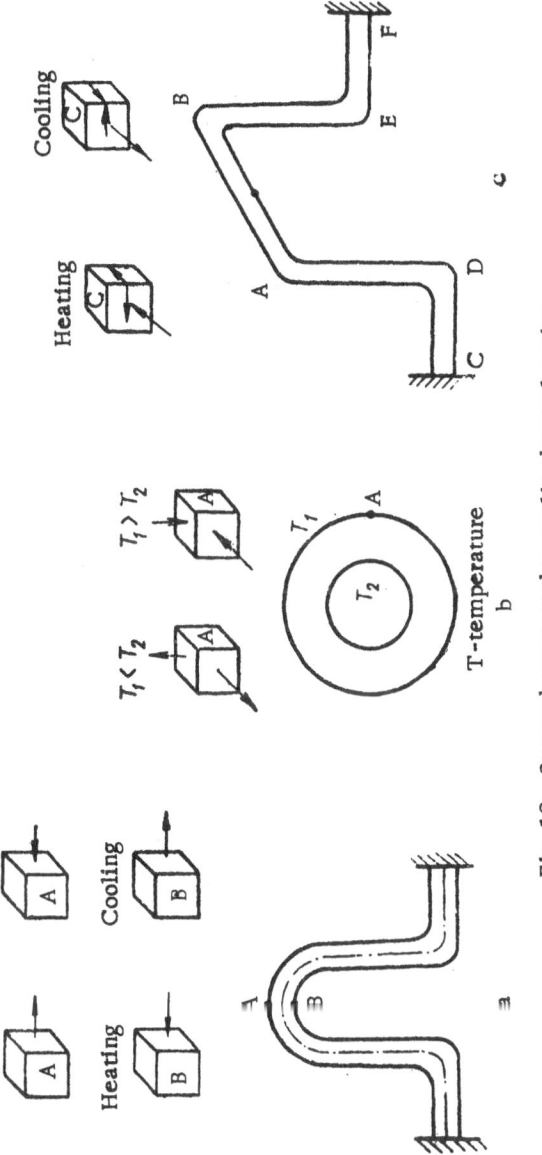

Fig. 12. Stressed states under cyclic thermal action.

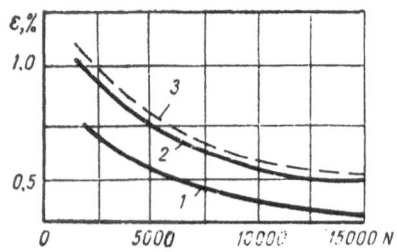

Fig. 13. Results of thermal fatigue tests under biaxial (curve 1) and uniaxial (curve 2) stressed states; curve 3 gives results of V. N. Kuznetsov's tests recalculated for the equivalent uniaxial stressed state.

Large thermal stresses occur in the pipe lines of energy conversion and chemical installations when there is a change in the temperature of the medium (vapor, gas, liquid) going through the lines.

In the space loop of the pipeline shown in Fig. 12c, the surface points of the segment AB show a complicated plane stressed state, which comes from deformation in tension and compression and from bending and torsion.

A volume stressed state is formed in the interior points of thick-walled parts under heating and cooling. The simplest example is provided by the triaxial tension at the center of a sphere when the surface layers are heated. Triaxial compression occurs at the center of the sphere on cooling.

The examples given show that uni-, bi-, and triaxially stressed and deformed states may be encountered under actual thermal effects.

This means that it is necessary to set up an engineering theory of strength. If we know even an approximate criterion of strength, we can find the behavior of the material in any stressed state from the results of tests in any one simpler stressed state. Only the first steps have been taken in this direction.

In the paper by V. N. Kuznetsov [16], a test was made under cyclic heating and cooling of tubular samples (6 × 1 mm) of 1Kh18N9T stainless steel. As a result of the radial temperature gradient, a plane compression stressed state was produced in the surface layers of the tube on cooling, and a plane tension stressed state was produced on heating. The difference in the amount of deformation was due to change in the radial temperature gradient.

The results of the tests are shown (Fig. 13, curve 1) in the form "elastoplastic deformation, ε, %, vs. number of cycles N to flaw formation." The magnitude of the deformation was found by calculation.

Figure 13 also gives the results of Coffin's tests [14] on the uniaxial stressed state for steel of similar composition (curve 2).

In order to compare his results with those of Coffin, V. N. Kuznetsov took the energy required to change the shape as the criterion of strength, and recalculated the results of his tests for the equivalent uniaxial stressed state (curve 3), making a correction for the reduced deformations.

It may be seen that curves 2 and 3 are quite close together in Fig. 13, which probably shows that there is some possibility of using the energy required to change the shape as a criterion of strength. It must, however, be kept in mind that in Coffin's experiments the maximum and minimum temperature drops were 100-600°C and 200-500°C, respectively, while in V. N. Kuznetsov's work, the temperature at the most highly stressed part varied from 20 to 150°C. This must be taken into consideration when comparing the results of the tests.

While V. N. Kuznetsov takes the criterion of strength to be the magnitude of the deformation, it may be seen from the results of Coffin's various tests (see Fig. 5), which are represented by the single curve "stress vs. number of cycles to failure," that using the stresses as a criterion of strength is possibly a good idea [23]. Further, some indirect inferences may be drawn as to the decisive effect of shear displacements.

At the present time, the experimental data are still too limited to make any definite recommendations as to what theory of strength to use. More quantitative studies of thermal fatigue need to be made for stressed states with different ratios of the principal stresses. In particular, to solve this problem, a method has been worked out for testing metals for thermal fatigue under conditions of pure shear [65].

Concentration of Stresses and Deformations in Thermal Fatigue

The majority of real parts have various types of cutouts and section changes such as openings, grooves, fillets, etc. Concentrations of stresses and deformations are known to occur at these points, and they may have a substantial effect on both mechanical and thermal fatigue.

TABLE 1. Depth of Flaws (in mm) as a Function of Slot Radius
in a Carbon Steel Sample [25]

Max. temperature of cycle, °C	Radius of slot, mm				
	0.15	0.28	0.52	1.02	4.0
600	1.0	0.6	0	0	0
700	6.4	4.1	3.5	1.5	0
800	10.6	6.6	4.9	2.3	0
900	6.2	6.1	3.9	0.7	0

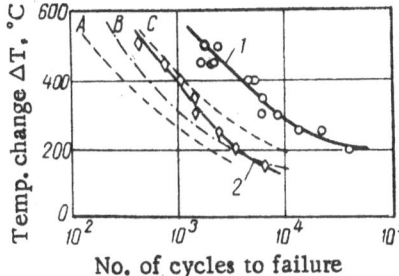

Fig. 14. Duration under temperature fatigue with stress concentration for 347 stainless steel [14]: (1) smooth sample; (2) sample with opening, d = 1 mm; (A, B, C) results of theoretical calculations by various methods.

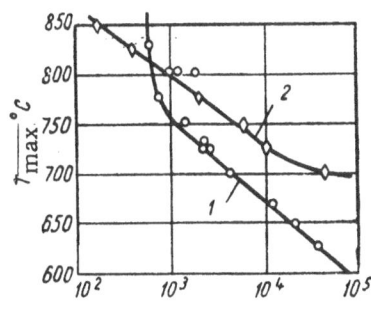

Fig. 15. Endurance at various maximum temperatures in the cycle, T_{min} 93°C [30]:(1) S-816; (2) Inconel 550.

It is important that under cyclic thermal action these points do not undergo purely mechanical concentration of stresses and deformations alone, but the nature of the temperature field changes to a considerable degree as compared with what is observed without the cutouts. This naturally renders it more difficult to make analytical calculations.

In the earlier work on thermal fatigue, the only thing that was investigated was the overall final result, without analyzing the stressed and deformed state. In L. A. Glikman's paper [24], a comparison was made between the number of cycles to failure for prismatic samples with and without longitudinal grooves. It was found that flaws show up more quickly in samples with grooves, for carbon steel after 150 cycles in samples with grooves and after 300 cycles in samples without grooves, for chrome-molybdenum steel after 250 cycles in samples with grooves, and flaws are completely absent in samples without notches.

Rädeker's study [25] on the effect of radius of rounding at the bottom of longitudinal grooves made in the surface of cylinders (Table 1) has shown that the depth of the flaws formed under cyclic heating and cooling increased with decrease in the radius, as a result of the large increase in stress concentration. The deepening of the flaws when the maximum temperature was raised from 600 to 800°C may be accounted for by the increase in thermal stresses. However, the reduction in depth of the flaws at $T_{max} = 900$°C is obviously due to relaxation of the stresses resulting from intensive occurrence of local plastic deformation.

Very dangerous places for stress concentrations are the welded joints between parts [8, 26, 27], because of the nonuniformity of the material at the welds, and the large residual stresses that are present.

Various types of roughness on the surface of parts, depending on previous machining, greatly reduce the resistance to thermal fatigue, since they act as stress concentrators [6, 11, 28, 29].

Coffin [14] has investigated the behavior of 347 stainless steel in thermal fatigue under conditions where there is stress concentration, by making tests on clamped tubular samples, with a circular hole 1 mm in diameter, under cyclic heating and cooling.

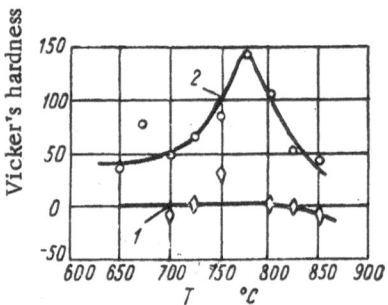

Fig. 16. Effect of T_{max} on the hardness change in thermal fatigue [30]: (1) Inconel 500; (2) S-816.

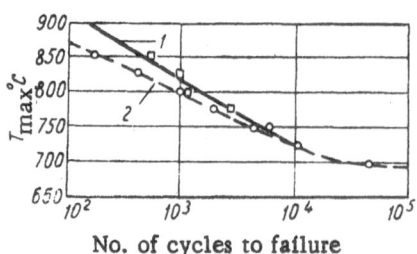

Fig. 17. Effect of T_{max} at constant temperature drop and at constant value of T_{min}. Duration of cycle 30 sec, held at T_{max} for 15 sec [21]: (1) ΔT (620°C = const); (2) $T_{min} \approx 95$°C.

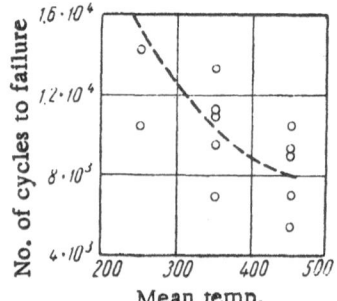

Fig. 18. Effect of mean temperature of cycle. Temperature drop 300°C [14].

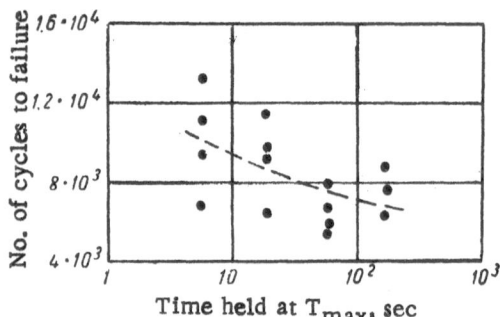

Fig. 19. Effect of time held at T_{max} with the temperature changing from 200 to 500°C. Material, 347 stainless steel [14].

The results of the tests are shown in Fig. 14 which gives the logarithm of the number of cycles to failure as a function of the magnitude of the cyclic temperature change for tubular samples with and without openings. It may be seen that both the experimental curves have the same slope, and that the effect of stress concentration is to shift the curve to the left, i.e., there is a reduction in endurance (at the same stresses), or a reduction in strength (for the same endurance).

A calculation giving a comparison of the behavior of a material with and without stress concentration is made below.

Thermal Fatigue as Affected by the Nature of the Temperature Cycle

Under the actual conditions of operation of a material, as well as when making tests, different relations are observed between temperature and mechanical cycles.

Two characteristic cases may be mentioned:

1. Under conditions where deformation is absolutely prevented, the mechanical deformation is wholly determined by the temperature drop. Thus, different temperature drops give different amounts of mechanical deformation.

2. Under definite conditions, the mechanical deformation can change over rather wide limits (for the same temperature drop in the element of volume in question). It will be shown below that this change depends on the stiffness of the elements in the system, and the nature of the temperature effect.

The majority of thermal fatigue studies have been made on a clamped sample. The fatigue curves obtained in this way represent a simultaneous change in mutually coupled factors — the temperature drop and the change in deformation. Here, although the magnitude of the total deformation is almost solely dependent on the temperature drop, the temperature level at which the thermal fatigue occurs is also important.

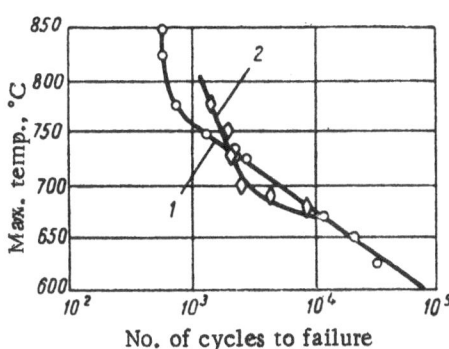

Fig. 20. Effect of time at T_{max}
for S-816 alloy [21]: (1) 15 sec; (2) 60 sec.

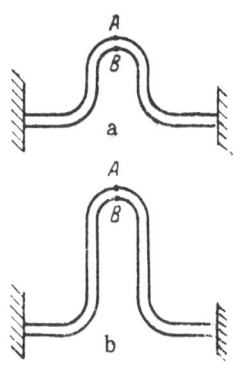

Fig. 21. Pipe
expansion loop

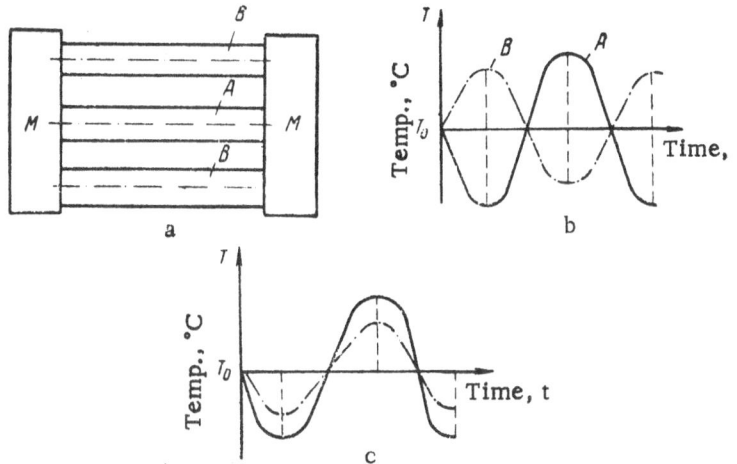

Fig. 22. Statically indeterminate system under thermal action.

Since, as the maximum temperature T_{max} of the cycle is increased, large changes in the structure and properties of the material occur, it is possible, on the one hand, to have localization of the deformation as a result of nonuniform heating while, on the other, some consideration must be given to the effect which the nature of the temperature cycle has on the behavior of the material in thermal fatigue.

Figure 15 gives the results of tests [30] on two materials: S-816 and Inconel 550. The minimum temperature of the cycle was constant and equal to 93°C. The maximum temperature of the cycle, T_{max}, was varied between wide limits. It may be seen from Fig. 15 that as T_{max} is increased, the number of cycles to failure drops considerably. But at a temperature above 790°C, the effect is very small for S-816, which is probably due to reduction in work hardening (Fig. 16). This reduction comes from processes that tend to increase the plasticity of the material.

In order to find the effect of change in T_{max} without changing the amplitude of the temperature cycle, Clauss and Freeman made some tests under these conditions (Fig. 17). If the number of cycles to failure were determined solely by the magnitude of the plastic deformation per cycle, the upper curve of Fig. 17 would be vertical.

In reality, as the maximum temperature of the cycle is increased, the number of cycles to failure decreases sharply, which shows that increasing the level of T_{max} has a large effect on thermal fatigue, since it increases other effects which lower the resistance to failure, as well as localizing the deformation as a result of loss in local strength.

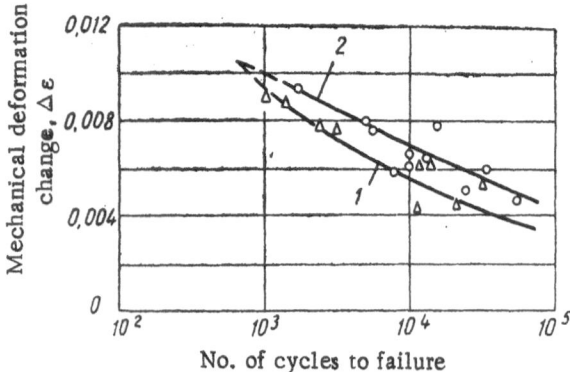

Fig. 23. Comparison of the endurance of a material with partially prevented deformation, and constant mean temperature, with the endurance for complete prevention of deformation and variable mean temperatures [23]: (1) partial clamping; (2) complete clamping.

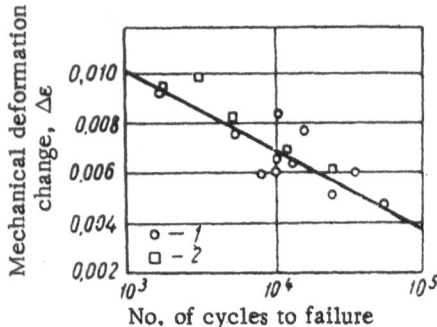

Fig. 24. Endurance of a material under thermal fatigue [23]: (1) at T_{max} — tension; (2) at T_{max} — compression.

It may also be seen from Fig. 18 [14] that the number of cycles to failure decreases as the mean temperature of the cycle is increased.

Increasing the time during which the material is held at the maximum temperature in the cycle increases the creep, as well as the effect which a number of factors have on the structure and properties.

Figure 19 shows the change in the number of cycles to failure with different lengths of time held at T_{max} for stainless steel [14]. The number of cycles to failure decreases as the length of time held at T_{max} becomes longer.

The results of tests on S-816 [21], where the material was held at T_{max} for 15 and 60 sec, show that the material behaves differently for different levels of T_{max} (Fig. 20). The increase in endurance when held at 60 sec as T_{max} is increased is due to processes which tend to increase the plasticity. The form of the curve for holding 60 sec at temperatures shows that these processes begin at a temperature of about 700°C.

It was noted above that under operating conditions, the mechanical deformation may be either greater or less than the thermal deformation. This is supported by the following examples.

For the two U-shaped pipe expansion loops (Fig. 21) with the same temperature excursion at the points A and B, the longer loop (see Fig. 21b) shows smaller linear deformations than the short loop (see Fig. 21a). This is accounted for by the differences in stiffness of the systems.

In Coffin's experiments on samples with rigidly clamped ends, a special case was mocked up for the operation of a volume element where the connected parts of the assemblies were absolutely rigid [14, 31].

Consider the statically indeterminate system consisting of the rod A and the two rods B, connected by rigid plates M (Fig. 22a). It is assumed that the rigidity of rod B is considerably greater than that of rod A.

Under temperature changes of the rod A alone, its mechanical deformation will be practically equal in absolute value to the temperature deformation, which corresponds with Coffin's early scheme. If, however, the rods A and B are simultaneously subjected to temperature changes, if the temperature change is as shown by the graph in Fig. 22b, the mechanical deformation of rod A will be greater than its thermal deformation, while if the temperature changes according to the graph of Fig. 22c, the mechanical deformation of rod A will be less than the thermal deformation. A similar picture will occur for a temperature change in rod A alone, if the stiffness of rod B is comparable with the stiffness of rod A.

Manson [1] feels that a sample clamped at the ends is an extreme case, and that the mechanical deformation is always less than the thermal deformation under actual conditions. However, it is impossible to agree with this from what has been said above.

It is important from both the practical and theoretical point of view to investigate the behavior of materials both when the mechanical deformations are greater than the thermal deformations and when the mechanical deformations are small.

In many practical cases, the systems are so yielding that the total elastic and plastic deformation is actually less than the thermal deformation.

The behavior of a material under these conditions has been studied experimentally by Coffin [23].

The study was made on the apparatus described below. Figure 23 shows the results of tests on 347 stainless steel. Curve 1 shows that the number of cycles to failure changes with change in mechanical deformation under conditions where the temperature cycle is constant (600 $\rightleftharpoons$ 100°C). Curve 2 is for a test on a sample with rigidly clamped ends, where the mechanical deformation changed as a result in change in T_{max} for a constant value of $T_{min} = 100$°C.

In the second case, for a fixed value of mechanical deformation, the number of cycles to failure was greater because of the lower value of the upper temperature of the cycle, as a result of the decisive role played by plastic deformation and thermal fatigue. If the quantity plotted as ordinate in Fig. 23 were not the total deformation but only the plastic part, one would expect the curves to come together, since the amount of plastic deformation increases at the higher value of T_{max} corresponding with curve 1.

It is obvious that the curves in Fig. 23 intersect at a point where the temperature cycles are the same in both types of test for the same deformation. To the left of this point (large deformations), the number of cycles to failure is less for tests on samples with clamped ends than for tests where the temperature cycles are fixed, since, in the first test, the same deformation may be found for a larger temperature drop, and hence for a larger value of T_{max}.

Plotting these same test results as "stress level vs. number of cycles to failure" (see Fig. 5) shows, as Coffin has pointed out, that there is no essential difference between the ways of fixing the mechanical deformation (by clamping the ends and changing the upper temperature of the cycle, or by producing mechanical deformations of various amounts with the temperature cycle constant).

It is possible that this fact indicates that a predominant role is being played in the fatigue process by the stress level, independent of the deformation level.

It was shown in [14] that first clamping the sample — at either the highest or lowest temperature in the cycle — has only a small effect on the rest of the deformation process. The difference is essentially only that in the first case the sample receives cold working in the first cycle from tension, while in the second case the cold working is produced by compression. The cold working is less in tension than in compression, since the deformation occurs at decreasing temperatures. In subsequent deformation in both cases, T_{max} acts at the instant of compression, while T_{min} acts at the instant of tension.

This difference has practically no effect on the number of cycles to failure. In tests on stainless steel [14], the number of cycles to failure for the same temperature cycle of 200 $\rightleftharpoons$ 500°C was 10,000 for the first method and 10,600 for the second.

There is also practical interest in the case where the tensile stress reaches its maximum value at T_{max}. Thus, in the pipe expansion loop (see Fig. 21), tension occurs at the points A as the temperature is raised, while compression occurs on cooling.

Similar experiments were made on an apparatus where the mechanical deformation was produced independently of the temperature cycle [23]. The results of these tests are given in Fig. 24, from which it may be seen that the sign of the deformation at the instant T_{max} is reached has no effect on the endurance of the material.

The results showed a considerable spread, as a result of the difficulty in holding the proper phase relation between the changes in temperature and mechanical deformation. Accordingly, the conclusion that the endurance is independent of the sign of the deformation must be regarded as a first approximation. But this fact obviously makes it possible to draw the conclusion that in thermal fatigue the decisive criterion of the deformed state is provided by the maximum shear deformations.

Accordingly, the greatest amount of damage to the material occurs at the instant T_{max} is acting, and this is where the comparison of the deformed states must be made. A comparison shows that the number of cycles to

TABLE 2. Limiting Values of Endurance of ÉI437 Alloy

Endurance limit	No. of cycles				
	10^6	10^7	$2 \cdot 10^7$	10^8	$5 \cdot 10^8$
$(\sigma_{-1})_N$, kg/mm² at 700°C	46	38	35	30.5	26
$(\sigma_{-1})_N$, kg/mm² at 800°C	33	27	25	21	11

TABLE 3. Limiting Values of Endurance of ÉI598 Alloy

Endurance limit	No. of cycles			
	10^6	10^7	$2 \cdot 10^7$	10^8
$(\sigma_{-1})_N$, kg/mm² at 800°C	32	26.5	25	17
$(\sigma_{-1})_N$, kg/mm² at 900°C	24	21	18.5	14

failure is the same in either uniaxial tension or uniaxial compression. This may be accounted for by the fact that the maximum displacements, which determine the damage to the material, are the same in both cases.

Mechanical Behavior of Materials under Cyclic Temperature Change

Under cyclic temperature changes, the mechanical behavior of the material is changing at every instant of time, and this affects the mechanical properties.

The conditions under which mechanical deformation occurs may be qualitatively different, depending on how long the temperature change cycle lasts. With a high frequency cycle, it is to be expected that failure will occur principally as a result of fatigue damage to the material. With a low-frequency cycle, the damage to the material will occur principally as a result of prolonged action of the stresses, i.e., failure will be determined by the long-time strength characteristics. In both cases, appreciable changes occur in the mechanical properties of the material if the temperature level and the nature of the temperature cycle are changed.

The data on the change in fatigue limits of a material with change in temperature level [33] provide an indirect indication that under rapid temperature change the residual damage buildup is different for each instant of time (and hence for each temperature). The limiting values of endurance for a number of structural alloys are given (Tables 2-4).

A direct proof of the effect of temperature change on fatigue damage is provided by the comparative experiments of Coffin [14] on stainless steel where, in one case, the samples were tested under cyclic temperature change from 100 to 600°C, with the corresponding cyclic deformation change, while in the other case the tests were made with cyclic change in deformation at a constant temperature of 350°C. It may be seen from Table 5 that the number of cycles to failure is about a third under cyclic temperature change of what it is at constant temperature. Here the magnitude of the deformation per cycle was the same in both cases and equal to 0.25 mm.

A proof of the fact that the amount of damage done to the material depends on the temperature level under slow temperature change is provided by the data of [32] on the change in long-time ultimate strength at different constant temperatures for 15Kh1MlF heat-resistant pearlitic tube steel used for piping in superheated steam power plants operating under steam conditions of 565-585°C and 140 atm pressure. The steel samples were subjected to normalization at 1050°C and soaking at 740°C for five hours.

The long-time ultimate strengths, found by extrapolation, were 8 kg/mm² at 570°C, 7.3 kg/mm² at 585°C, and 6.1 kg/mm² at 600°C.

A direct proof of the effect of temperature change on the damage done under prolonged stress is found in the data of R. N. Sizova and N. Ya. Nikolenko, shown in Fig. 25, where the long-time strength curve for the cyclic temperature change 750-850-750°C falls below the long-time strength curve for the mean temperature of 800°C.

TABLE 4. Limiting Values of Endurance of 40KhNMA Steel

Endurance limit	No. of cycles				
	10^6	$3 \cdot 10^6$	10^7	$1.8 \cdot 10^7$	10^8
$(\sigma_{-1})_N$, kg/mm² at 20°C	63	55	55	55	55
$(\sigma_{-1})_N$, kg/mm² at 300°C	60	52.5	44	40	40
$(\sigma_{-1})_N$, kg/mm² at 550°C	44	38	31	28	20

TABLE 5. Comparison Between the Number of Cycles to Failure
for Cyclic Change of Temperature and Deformation [14]

State of sample	Number of cycles to failure	
	Cyclic temperature change from 100 to 600°C	Cyclic deformation change at 350°C
Annealed	1760,2260	5820,5950
Accumulated (γ = 0.816)	650,880	2130,2400

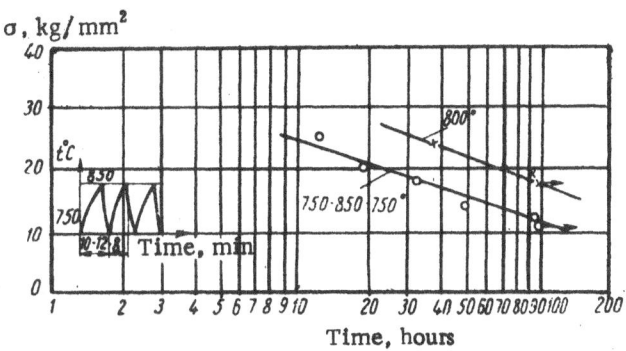

Fig. 25. Long-time strength curves of ÉI437 alloy (d_{rest} = 3.4 mm) at a constant temperature of 800°C, and with the cyclic temperature change 750-850-750°C.

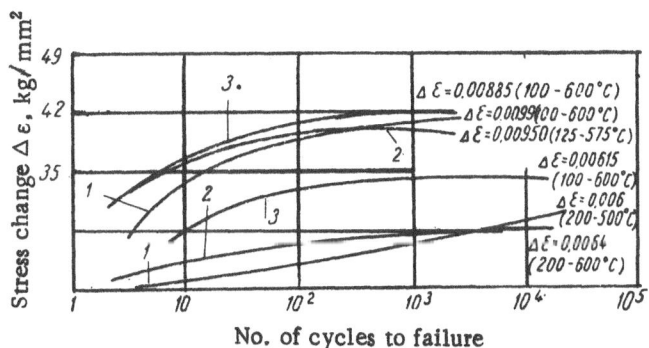

Fig. 26. Stress change for different cyclic temperature changes [23]: (1) complete clamping — compression at T_{max}; (2) complete clamping — tension at T_{max}; (3) partial clamping — tension at T_{max}.

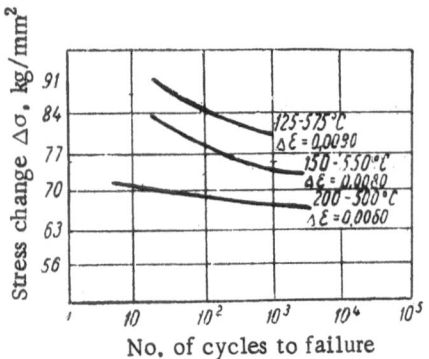

Fig. 27. Stress change under cyclic temperature change (180° torsion, with a calculated length of 50 mm) [14].

The tests of [14, 23] have shown that the resistance to plastic deformation changes during thermal fatigue. This affects the way the stresses change with time for a given amplitude of deformation. While work hardening occurs in an annealed material with increase in the number of cycles (Fig. 26), a material that is already worked shows weakening (Fig. 27). It may be seen from Fig. 26 that the same law holds for different forms of the cycle.

In considering deformation under thermal fatigue, attention must be given to the way in which the resistance to plastic deformation varies with the heating and cooling rate.

Actually, a large enough amount of heat exchange can produce a rapid change in the temperature gradient, which increases the rate of increase of the stresses. It has been shown in the work of F. F. Vitman and N. A. Zlatin [34, 35], and of Nadai [36], that the deformation rate is essentially dependent on the resistance to plastic deformation. Accordingly, as the heat exchange rate is increased, an increase may occur in the acting stresses.

It has been shown by the calculations of [2] that in the initial period of cooling the surface layers of the cylinder, the elastic deformation rate is about 0.5% per second. If the flow limit is passed, the deformation rate will be lower (0.1-0.01% per second). But even a rate this low can exert an effect on the process.

Effect of Previous Working on the Thermal Fatigue Resistance

There is a large amount of theoretical and practical interest in finding what effect previous plastic deformation has on the thermal fatigue resistance.

Coffin [14] made an experimental study of the effect of working on thermal fatigue for 347 stainless steel (Fig. 28).

Although in mechanical fatigue tests previous deformation almost always increases the number of cycles to failure, it may be seen from Fig. 28 that when cyclic thermal stresses are acting, the strength of the worked material drops off if the magnitude of the deformation per cycle is relatively large. This is obviously due to the fact that in making mechanical fatigue tests the effect of working is usually evaluated for fixed stresses, while in thermal fatigue it is the deformations that are given.

Working increases the flow limit so that, for fixed stresses, the amount of plastic deformation drops off, with the result that there is an increase in the mechanical fatigue strength. On the other hand, if the magnitude of the deformation is fixed, the reduction in plasticity from previous working produces an appreciable increase in the stresses (the amount of plastic deformation changes very little for practical purposes) and a reduction in the number of cycles to failure (Fig. 29).

Coffin has also shown that previous working may be useful in thermal fatigue at relatively small deformations, i.e., with a relatively small temperature drop ΔT. This is due to the fact that for small fixed deformations in a worked material with a high flow limit, there will be no regular plastic deformations to affect the fatigue damage (Fig. 30).

There are a number of other papers [6, 37] where the effect of working has been investigated. In making a study of the experimental results in these papers, account must be taken of the effect of the residual stresses produced from previous plastic deformation.

V. M. Stepanov [37] has made a thermal fatigue study of 1.5-mm thick sheet materials made of austenitic steel and nickel alloys. Part of the samples were subjected to working by bending the sample to 90° and straightening them out again.

The tests made have shown that the number of cycles to first appearance of flaws is less in the cold-worked samples than in the annealed samples by 10 to 80%. In this case, the negative effect of cold working is enhanced by having residual tensile stresses present.

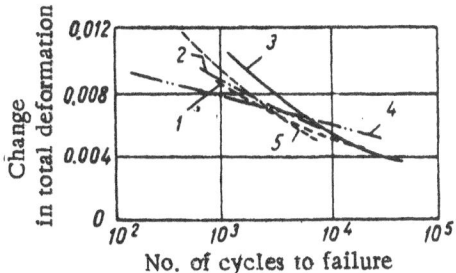

Fig. 28. Effect of previous working on the thermal fatigue resistance [14] $\Delta \varepsilon_T$ = $\Delta(\varepsilon_y + \varepsilon_p) = \alpha(\Delta T)$: (1) 30% tension; (2) 180° torsion; (3) annealing; (4) 360° torsion; (5) 15% tension.

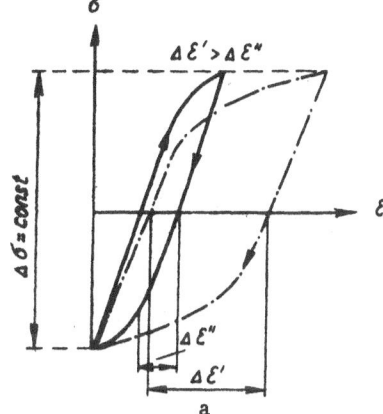

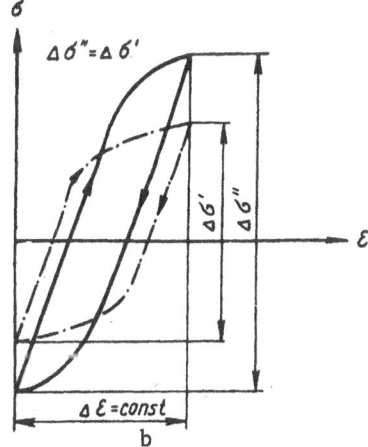

Fig. 29. Nature of the deformation in worked and annealed materials: (a) stress change fixed; (b) deformation change fixed.

The tests made have shown that the number of cycles to first appearance of flaws is less in the cold-worked samples than in the annealed samples by 10 to 80%. In this case, the negative effect of cold working is enhanced by having residual tensile stresses present.

D. I. Kostenko [6] observed an increase of 20% in the number of cycles to appearance of thermal fatigue flaws in samples subjected to shot blasting.

Shot blasting the surface of punches also increased their service life 20-50%. The author explains the favorable effect of cold working on the behavior of materials under thermal fatigue by saying that cold working with shot flattens out or removes the surface defects, and produces residual compressive stresses and strengthens the surface layers of the metal.

A study of the effect of previous prolonged loading, which produces cold working in long-time strength tests has been made for S-816 alloy and Inconel 550 by Clauss and Freeman [21] (Figs. 31 and 32).

It may be seen that in S-816 alloy the thermal fatigue endurance is reduced if the time kept under load is increased, while no appreciable effect was observed for Inconel 550.

Manson [1] has attempted to make an analytical calculation of the effect of previously keeping a material under stress at elevated temperature. It is assumed as a first approximation that the change in strength and thermal fatigue is caused by the change in plasticity of the material from keeping it under stress.

Thus, it is only necessary to measure the remaining plasticity and tension, and use Eq. (6) to calculate the endurance of the material. If Eq. (6) is used for tension tests, we obtain

$$D^n N_0 = k, \tag{7}$$

where D is the plasticity of the material in tension, and N_0 is the equivalent number of cycles to failure.

For two samples held under load for different times, the plasticity and tension will be D_1 and D_2, respectively.

The endurances, N_1 and N_2, of these samples, tested for thermal fatigue under the same conditions, will be related by the equation

$$\frac{N_1}{N_2} = \left(\frac{D_2}{D_1}\right)^n. \tag{8}$$

It is assumed here that the amount of plastic deformation per cycle is directly proportional to the plasticity in tension. Thus, if D_1 and N_1 are known, N_2 may be calculated by determining D_2. The temperature to be used when finding the plasticity requires special attention. It is assumed that the decisive value is the maximum temperature of the cycle.

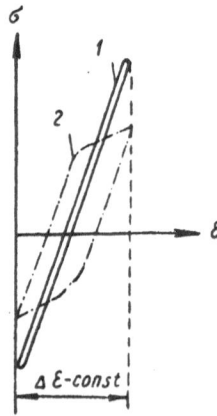

Fig. 30. Deformation of cold-worked and annealed materials at small deformations: (1) cold-worked material; (2) annealed material.

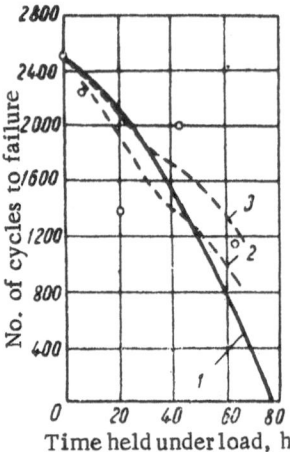

Fig. 31. Effect of prolonged loading on thermal fatigue of S-816 alloy. Temperature cycle 97 → 730°C. Long-time loading: σ = 28 kg per mm², 730°C [21]: (1) experimental data; (2) data calculated for n = 3; (3) data calculated for n = 2.

The dotted lines in Figs. 31 and 32 give the results of analytic calculations from Eq. (8). Although these results give some measure of justification for using Eq. (8), additional investigations must be made to arrive at a final solution.*

A problem no less important than thermal fatigue is that of the behavior of a material with nonuniform distribution of the properties that go to make up the resistance to plastic deformation.

Coffin [14] has made tests on samples that had been cold worked over only a part of a definite length. The samples were subjected to cyclic plastic deformation at constant temperature. The reduction in the ratio a (Fig. 33) produced a reduction in endurance, due to an increase in concentration of the deformation at the annealed part with a constant total deformation.

It may be seen from Fig. 33 that the samples with the ratio a = 0 should have had an endurance of about 200 cycles, while the samples that were uniformly cold worked over the whole length show endurances of up to 2000 cycles. This may be accounted for as follows: extrapolating the experimental curve to the value a = 0 assumes that the unworked portion remains of very small length, and that localization of the deformation occurs there. Although the experiment was made at constant temperature, the basic results may be extended to the case of cyclic temperature change.

Thermal Fatigue as Affected by Mechanical and Thermal Properties of the Material

In the majority of thermal fatigue studies, no analysis was made of the temperature field, and hence of the stress and deformation fields, which bear a direct relation to the heat transfer and heat conduction. Accordingly, a comparison of the behavior of different materials as a function of their mechanical properties can only be made in the case where samples of the same shape are being tested under approximately identical conditions of heat transfer and heat conduction.

A study of the effect of hardness on thermal fatigue was made by D. I. Kostenko [28] for 5KhNT and 5KhNM die steel (Fig. 34). As the hardness increased, the number of cycles to failure dropped off, due to reduction in plasticity, as well as partly to the samples with high hardness having a martensite or troostite structure which has lower thermal conductivity than the sorbite structure.

It is shown in the paper by Thielsch [8] that for cobalt alloys tested under the same conditions, the number of cycles to failure is linear when plotted semilogarithmically against $\alpha / \lambda \varepsilon$, where α is the coefficient of linear expansion, and λ is the coefficient of heat conduction, which remain practically constant for these alloys. It was found that the number of cycles to failure decreased with decrease in the plasticity ε.

*The plasticity D for different times held under load is found from the results of the tests shown in Fig. 41.

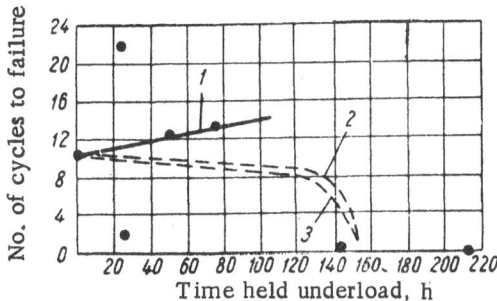

Fig. 32. Effect of prolonged loading on thermal fatigue of Inconel alloy 550. Temperature cycle 95 → 730°C. Prolonged loading: $\sigma = 40$ kg/mm^2, 750° [21]: (1) experimental data; (2) data calculated for n = 2; (3) data calculated for n = 3.

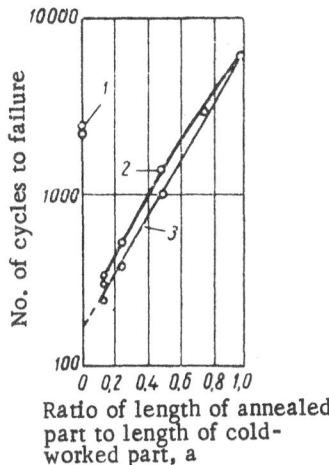

Fig. 33. Results of thermal fatigue tests on 347 stainless steel samples with nonuniform cold working along the length [14]: $T_m = 350$°C; $\varepsilon = 1\%$: (1) uniformly cold-worked sample; (2) data calculated for nonuniformly cold-worked sample; (3) experimental data for nonuniformly cold-worked sample.

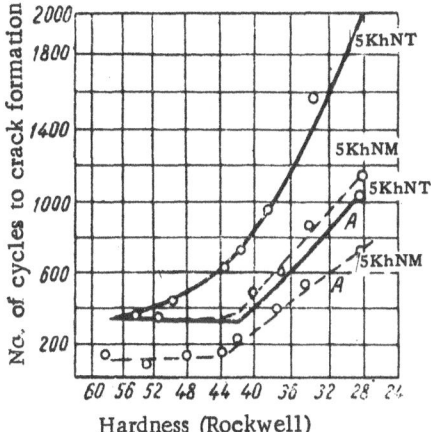

Fig. 34. Effect of hardness of 5KhNT and 5KhNM steel on the formation of longitudinal thermal fatigue cracks (average of data from 3-6 tests) [28]: (A) start of crack formation; (B) crack formation over the whole length.

V. I. Zalesskii and D. M. Korneev [4] made tests on 4KhNV and 4KhVS steels, which have approximately the same static characteristics $\sigma_{0.2}$, σ_B, and δ.

When tested under the same conditions, the first steel failed after 120 cycles, and the second failed after 260 cycles.

The difference in thermal fatigue may be accounted for as a difference in maximum endurance, if it is assumed that the stressed state was the same in both cases.

An example of studies made for the purpose of finding the behavior of different materials for fixed ranges of change in the temperature of the surrounding medium is provided by the work of Hunter [38].

The results of this work are shown in Table 6. Since there was no analysis of the stressed state, which is determined by the difference in heat conduction and heat transfer conditions, these results are only of limited practical interest, as has been correctly pointed out by Yu. F. Balandin [3].

The thermal characteristics — the coefficient of heat conduction λ, the coefficient of linear expansion α, and the heat transfer coefficient h — have a substantial effect on the behavior of the material under cyclically varying temperatures.

The attempt made in [39] to calculate the resistance of the material to thermal fatigue, including the change in thermal characteristics, but neglecting the change in mechanical properties, did not result in any useful relations, since, as a rule, when the thermal characteristics change, the mechanical behavior of the material changes also.

How to make a quantitative treatment of the simultaneous effect of thermal and mechanical properties with plastic deformations present is a very pertinent question, but is a rather difficult problem. To a first

TABLE 6. Number of Cycles to Appearance of Cracks in Samples of Various Materials [38]

Material	Maximum temperature of cycle, °C				
	870	930	980	1040	1100
Batelle alloy (chromium base)	12,000	12,000	7400	3850	3820
S-816 Kovan (cobalt base)	4200	1820	990	1170	890
Waspalloy (nickel base)	12,500	1160	1060	1060	890
Hastelloy C (nickel base)	4800	3100	2040	3700	330
M-252 (nickel base)	11,700	8000	2220	1760	1330
K-152B (cermet)	8500	—	390	—	250
347 SS (18-8 stainless)	3000	2000	—	—	—
310 SS (25-20 stainless)	6500	3800	1600	1070	750
Niconel B (nickel base)	3200	2350	1290	470	390

TABLE 7. Stress Calculations for Various Materials [2]

Characteristic of material	45 carbon steel	ÉI437 heat-resistant alloy	ÉI69 austenitic stainless steel
$E \cdot 10^6$, kg/mm^2	2.06	1.99	1.85
$a \cdot 10^{-6}$, 1/°C	12	12.67	18
Ea, kg/mm$^2 \cdot$ deg	24.7	25.4	33
V*, cal/cm$^2 \cdot$ sec $\cdot$ deg	0.05	0.05	0.05
λ, cal/cm $\cdot$ sec $\cdot$ deg	0.13	0.046	0.04
R, cm	5	5	5
Rh	19	5.4	6.25
ψ_{max}	0.25	0.4	0.415
ΔT, °C	600	600	600
σ_θ, kg/mm^2	53	86	118
σ_s, kg/mm^2	36	56	47
σ_θ / σ_s	1.47	1.53	2.5

approximation, this problem may be solved for an elastic stress distribution using the relations obtained by Manson [1] in calculating the resistance of brittle materials to thermal shock.

In the paper by S. V. Serensen and P. I. Kotov [2], a calculation is made for the thermoelastic stresses occurring in cylinders of a given radius made of different materials for the same temperature drop.

It may be seen from Table 7 that in a less heat-conductive material, such as ÉI69 austenitic steel, the (calculated) thermal stresses that occur are greater, other conditions being equal, than in the more heat-conductive 45 carbon steel.

These calculations permit a qualitative comparison to be made of the behavior of the materials in thermal fatigue.

Thermal Fatigue as Affected by Different Factors

Such processes as aging, corrosion, and change in grain size can exert a substantial influence on the mechanical properties of a material, and hence on the way the material behaves during thermal fatigue.

Aging is of great importance, since the majority of the heat-resistant alloys used are in an unstable state. High temperature and plastic deformation favor transition of the structure to a stable state. The constituent parts of a solid solution have a tendency to precipitate, which can have a substantial effect on the mechanical properties. Thus, precipitation at the grain faces reduces the plasticity, which leads to more rapid failure under cyclic loading.

The surfaces of parts operating under cyclic temperatures are usually in contact with oxygen and other gases. At high temperatures, reaction with the surrounding medium produces brittle films on the surface which,

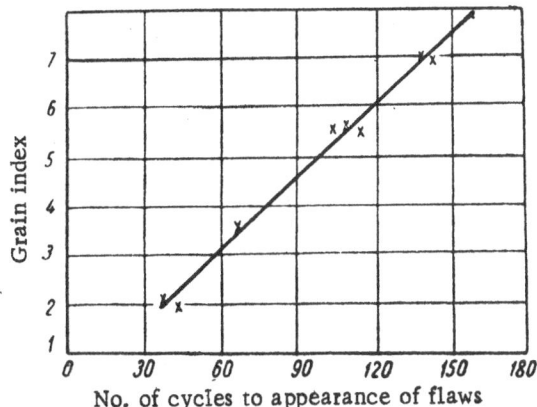

Fig. 35. Thermal fatigue of Kh20N80 alloy
as a function of grain size [40].

when broken, help to form cracks. Hydrogen from the ambient medium can diffuse into the grain boundaries, thus lowering the thermal fatigue resistance.

Some of the elements in nuclear reactors are subjected to the corrosive action of various coolants. Under cyclic deformation at elevated temperatures, this effect may be enhanced, and sometimes general corrosion changes to local corrosion, which produces rapid failure of the material.

Recrystallization of the material often occurs at elevated temperatures, which affects the resistance to thermal fatigue.

The data on the effect of grain size are rather contradictory. In Kh20N80 alloy [40], the number of cycles to failure increases with decrease in grain size (Fig. 35). At the same time, a reduction in thermal fatigue resistance was observed with decrease in grain size in thermal fatigue tests on one of the nickel-based alloys [41].

In further studies on the effect of grain dimension on thermal fatigue, consideration must be given to both stresses of the first kind and microstructure stresses of the second kind.

Coating parts may change the thermal fatigue resistance. Nickel plating, case-hardening, and copper plating increase the thermal fatigue resistance. Chromium plating and carburization have a bad effect [6, 24, 29, 39, 42-44].

The difference between the effects of protective coatings is due to the complicated relation between the thermal properties and the strength characteristics of the base metal and the coating.

It has been pointed out by Yu. F. Balandin [3] that a denser material, in which internal voids, pores, gas bubbles, and nonmetallic inclusions, etc., have been reduced to a minimum, should show better resistance to thermal fatigue. Thus, for example, forged materials have higher resistance that cast materials [8, 29, 45].

There is a series of papers dealing with the effect on thermal fatigue of impurities present in alloys and steels. For example, cracks develop more rapidly in steel if the silicon content is increased [5, 41, 46], and a bad effect is exerted by increasing the amount of magnesium, or having phosphorous present [5, 47].

S. T. Kishkin and A. A. Klypin [48] have shown that steels containing a large amount of carbon have inferior mechanical properties to steels with less carbon, after a definite number of thermal cycles.

Easily fusible impurities in heat-resistant alloys have a bad effect [48, 49]. If the thermostructural stresses are to be reduced, it is better to use alloys consisting of a single phase. But sometimes having two phases exerts a positive effect, if we consider the sum total of all the different factors, for example greater strength, increase in plasticity, etc. Thus, it is felt to be a good thing to have small quantities of ferrite present in cast austenitic steel [8, 42].

Deterioration of the Mechanical Properties of Materials during Thermal Fatigue
(Damage)

Complete failure from thermal fatigue is preceded by the formation of microscopic cracks. The cracks get larger with time, and join together, finally producing complete failure. After the first cracks are formed, the conditions under which they develop may change as a result of change in heating (for example, in heating by an electric current), or of change in heat transfer, etc.

Thus, it is stated in some papers that cracks stop penetrating into the samples for a while after a definite number of cycles [43]. In other papers [16, 50] it is pointed out that the cracks pass completely through the thickness of tube walls.

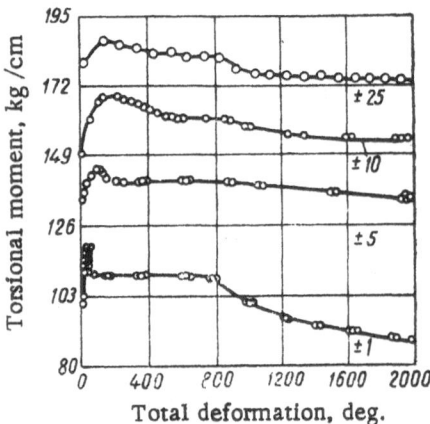

Fig. 36. Change in the flow resistance of copper samples under alternating plastic deformations [55] (the numbers on the curves are the deformations in degrees per cycle).

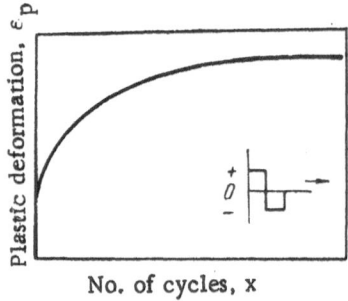

Fig. 37. Change in amount of plastic deformation per cycle with increase in the number of cycles. Total deformation constant [20].

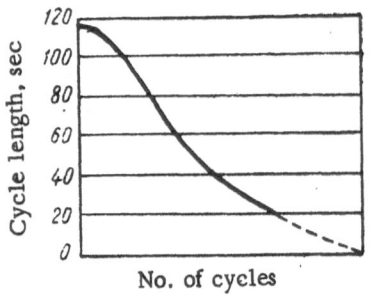

Fig. 38. Endurance of a material as a function of cycle length with constant plastic deformation [20].

A. V. Ratner [51] has made thermal fatigue tests on 15KhM steel tubes. The corrosion and fatigue cracks formed at weak points changed their depth very little as the number of cycles was increased. The author accounts for this by saying that as the cracks penetrate into the tube and the number of surface cracks increases, the thermal stresses at the bottom of the cracks decrease considerably, and if there are no large alternating mechanical stresses, the cracks may fail to penetrate the entire thickness of the tube wall. This explanation is obviously plausible since, even in mechanical fatigue, there are definite conditions of loading under which crack development may stop [52]. One of the factors that helps to stop thermal fatigue cracks is that the thermal stresses are taken off of the surface layers when the cracks form.

The way in which the thermal fatigue cracks develop has been very little studied. In some steels and alloys under cyclically varying temperatures, the cracks develop both in the grains and along the grain boundaries, while in others they only develop along the boundaries [53, 54].

The damage done to the material in thermal fatigue shows up either as a change in resistance to plastic deformation or as a change in resistance to failure.

Wood and Segall [55] tested copper samples previously cold-worked in tension (up to $\psi = 70\%$) under alternating deformation of constant amplitude but opposite sign. The change in resistance to plastic deformation after a definite number of cycles was noted from the torsional moment corresponding with the beginning of flow (Fig. 36). It may be seen that in all cases at the start of the cyclic temperature change, some increase occurred in the flow limit, and then as the number of cycles was increased, the flow resistance dropped off regularly, showing that the material had been weakened.

Kennedy's experiments [20] on Inconel 550 at elevated temperature have shown that for constant total deformation, the amount of plastic deformation increases with increase in the number of cycles (Fig. 37). It was also shown that for a constant load the time required to pass through a given amount of plastic deformation decreased continuously with increase in the number of cycles (Fig. 38). This shows that Inconel 550 is weakened by alternating deformation.

Everything that has been said makes it clear that cyclic thermal stresses can reduce the resistance to plastic deformation or creep.

The results of a number of studies show that cyclic temperature change has an appreciable effect on the long-time strength of a material.

Clauss and Freeman [21] have shown (Fig. 39) that the long-time strength of Inconel 550 drops off with increase in the number of thermal cycles, the dropoff becoming catastrophic

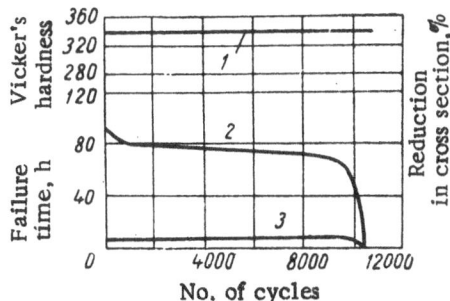

Fig. 39. Effect of thermal stresses on hardness and long-time strength of Inconel 550. Temperature cycle, 95-730°C. Long-time strength, $\sigma = 40$ kg/mm² at 730°C [21]: (1) hardness; (2) long-time strength; (3) reduction in cross section.

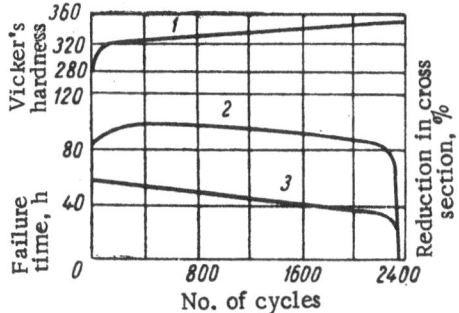

Fig. 40. Effect of thermal stresses on hardness and long-time strength of S-816 alloy. Temperature cycle, 95-730°C. Long-time strength, $\sigma = 28$ kg/mm² at 730°C [21]: (1) hardness; (2) long-time strength; (3) reduction in cross section.

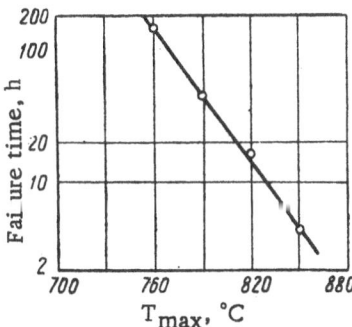

Fig. 41. Effect of T_{max} on the long-time strength of Inconel 550. $T_{min} = 95$°C. Long-time strength, $\sigma = 40$ kg/mm² at 730°C.

after a certain number of cycles. In S-816 alloy, the long-time strength increases after a certain number of cycles, but then also drops off (Fig. 40). The long-time strength of Inconel 550 drops off with increase in the maximum temperature of the cycle, T_{max} (Fig. 41). In these experiments, the samples were tested after a number of cycles equal to half the number of cycles required to produce failure. It is obvious that the drop-off is due to weakening processes and localization of the deformation.

Coffin [14] found that the first cracks appear quite rapidly in thermal fatigue, but that further growth is relatively slow. The amount of damage done to the material was evaluated by constructing curves of deformation and tension at normal temperature, after different numbers of thermal cycles.

It was found that the cracks formed produce an early break, since they serve to concentrate the stress. Further, the effect of cyclic deformation is to raise the deformation curve slightly, as a result of cold working of the material. It might be thought that the rest of the unfailed material would be undamaged, but there is a reduction in plasticity as a result of fatigue cracks. There is a rather large spread in the values of plasticity (Fig. 42).

The paper by S. T. Kishkin and A. A. Klypin [48] shows the effect of cyclic heating and cooling on the mechanical properties of various materials. Thus, short-time and long-time tests on 1Kh18N9T steel after a definite number of thermal cycles show that there is a reduction in strength (Fig. 43).

A comparison of the test results on different materials shows that the change in mechanical properties depends on the nature of the alloy, the cooling rate, the temperature to which it is heated, and the number of thermal cycles. It is also found that the higher the strength of the material, the less the reduction in resistance to failure.

The damage done to 1Kh18N9T steel in cyclic heating and cooling has been investigated by Ya. B. Fridman and V. I. Egorov [56]. It was shown that there is a large reduction in strength and plasticity from cyclic thermal stresses in both cold-worked and annealed materials (Figs. 44, 45). However, holding the material at the highest temperature has no material effect on the mechanical properties (Fig. 46). Stress concentration greatly reduces the strength of 1Kh18N9T steel after thermal action (Fig. 47).

Microstructural Stresses (of the Second Kind) in Thermal Fatigue

Microstructural thermal stresses (of the second kind) have an importance of their own since they can occur even when no stresses of the first kind are present.

It has been pointed out in many papers [2, 57, 58] that it is necessary to pay attention to the effect of thermal stresses

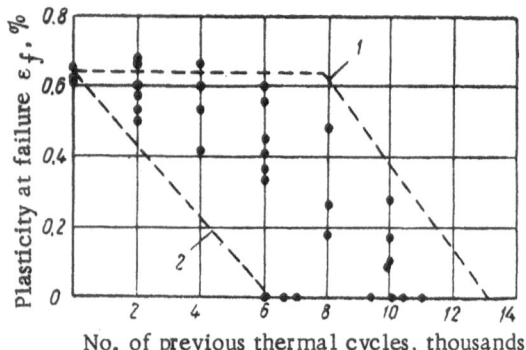

Fig. 42. Plasticity in tension as a function of the number of previous thermal cycles for 347 stainless steel [14]: (1) least damage; (2) greatest damage.

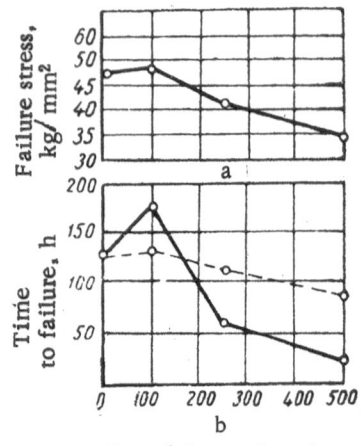

Fig. 43. Change in short-time and long-time strength of 1Kh18N9T steel at T = 800°C as a function of the number of thermal cycles (T_{max} = 800°C). Solid line: water cooling; dotted line: cooling in an air stream [48]: (a) short-time tests; (b) long-time tests.

of the second kind. It was shown in the paper by V.A. Likhachev [58] that thermostructural stresses are produced in a changing temperature field by the following causes:

a. anisotropic thermal expansion of at least one of the phases in heterogeneous systems;

b. anisotropic coefficient of thermal expansion of metals with a noncubic crystal lattice;

c. difference in the coefficients of thermal expansion of neighboring phases in heterogeneous systems; and,

d. phase transformations producing a change in specific volume of the phases.

Hence, it is clear that thermostructural stresses can occur even in a uniform temperature field with no temperature gradient in the body and no external mechanical restraints applied.

The magnitude and sign of the thermostructural stresses is affected by anisotropy in the microstructure, and by microstructural inhomogeneity in the mechanical properties of the material.

The presence of thermostructural stresses resulting from an anisotropic coefficient of thermal expansion has been shown by the experiments of Boas and Honeycombe [57] on pure metals (zinc, tin, cadmium) and alloys in the temperature range 30-150°C with uniform heating and cooling. These experiments showed plastic deformation in individual grains, which increased rapidly from cycle to cycle.

V. A. Likhachev [58], in investigating the effect of thermostructural stresses on irreversible change in shape, made experiments on cadmium (hexagonal lattice) and tin (tetragonal lattice), as these materials show anisotropic thermal expansion. In order to exclude any possible effect of stresses due to a temperature gradient, the cyclic temperature change was set up in such a way that there was practically no temperature gradient over the cross section. The results of these experiments have shown that if there is a grain structure in the samples, the thermostructural stresses show up as a change in dimensions resulting from stress relaxation.

The conclusion may be drawn from this work that for a large enough number of temperature cycles the accumulated damage from thermostructural stresses ultimately produces flaws in individual grains or along the grain boundaries, which then lead to complete failure. This type of failure may be regarded as thermal fatigue resulting from thermostructural stresses alone. Failure of this type under laboratory conditions means setting up very prolonged experiments, since the thermal cycle must take place with practically no temperature gradient over the cross section of the sample.

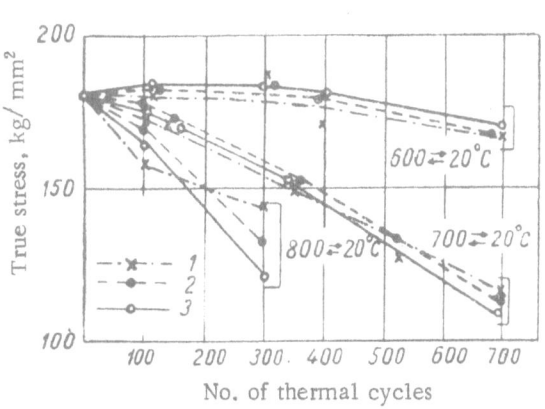

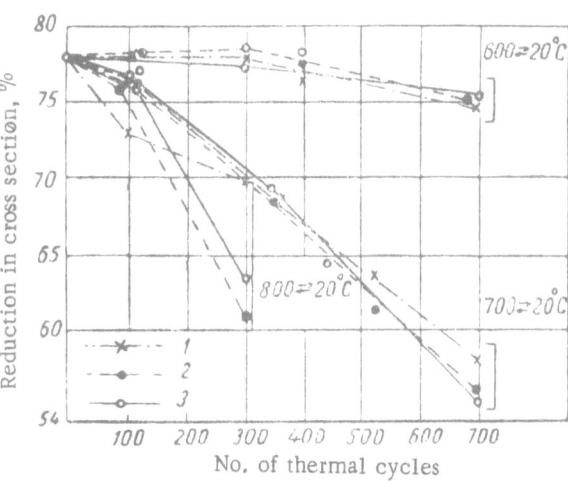

Fig. 44. True stress of 1Kh18N9T steel as a function of number of thermal cycles: (1) without previous deformation; (2) 5% previous deformation; (3) 20% previous deformation.

Fig. 45. Plasticity of 1Kh18N9T steel as a function of number of thermal cycles: (1) without previous deformation; (2) 5% previous deformation; (3) 20% previous deformation.

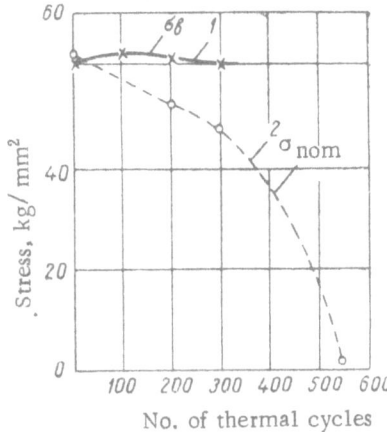

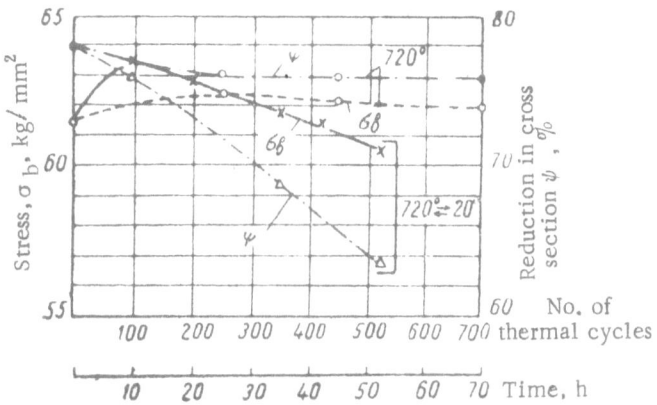

Fig. 46. Effect of stress concentration on damage done to 1Kh18N9T steel in thermal cycling: (1) sample without opening; (2) sample with opening.

Fig. 47. Mechanical properties of 1Kh18N9T steel as a function of number of thermal cycles (720 ⇌ 20°C) and of time held at 720°C.

The experiments of V. A. Likhachev [58] have shown that change in shape of samples as a result of thermo-structural stress relaxation is only observed when the material has a grain structure. The fact that there is no external effect in a material that has no grain structure is due to the statistically uniform distribution of the crystallites in the solid, although it is possible to have local change of shape in individual groups of crystallites.

As a result of this, thermal fatigue may be observed in either the presence or absence of grain structure, since thermostructural stresses can lead to failure in either case.

It was shown in [58] that increasing the grain size produces a considerable reduction of the growth coefficient in change in shape produced by anisotropic thermal expansion. It is thus to be expected that the endurance in thermal fatigue from thermostructural stresses will increase with increase in grain size.

The paper by Boas and Honeycombe [57] gives an equation for making an approximate calculation of the magnitude of the microstructural stresses. This equation is obtained from a consideration of the simplest system consisting of two rigidly bound elements of the same cross section but with different coefficients of linear

TABLE 8. Thermostructural Stress Parameters for Zinc and Cadmium

Metal	kg per $mm^2 \cdot deg$	kg per $mm^2 \cdot deg$	kg per $mm^2 \cdot deg$	kg per $mm^2 \cdot deg$	kg per $mm^2 \cdot deg$	kg per $mm^2 \cdot deg$
Cadmium	0.0467	0.0714	0.0253	0.0645	0.0108	0.00222
Zinc	0.0894	0.135	0.0520	0.125	0.0166	0.00482
$\sigma_i^{Zn}/\sigma_i^{Cd}$	1.94	1.91	1.89	2.05	1.54	2.17

expansion, α_1 and α_2, and different moduli of elasticity, E_1 and E_2, with a temperature change of ΔT °C:

$$\sigma = (\alpha_1 - \alpha_2)\,\Delta T\,\frac{E_1 E_2}{E_1 + E_2}\,. \tag{9}$$

It may be seen from Eq. (9) that the decisive factor in stress formation is the difference between the coefficients of linear expansion of the structural elements in some direction.

V. A. Likhachev [58], in making an approximate calculation of thermostructural stresses, has considered a model where, in the center of an isotropic sphere of radius R_0 there is a spherical inclusion of radius R, which has the same elastic properties as the sphere but a different coefficient of thermal expansion. From elasticity theory, the stresses were calculated in the inclusion when the body was heated by the amount ΔT °C with $R_0 = \infty$

$$\sigma_{ii}^b = \frac{(\sigma_{ii}^b)_0}{3\left(1 + \dfrac{3}{4}\dfrac{K}{G}\right)}\,, \tag{10}$$

where σ_{ii}^b are the stresses developed in the inclusion on heating, when it is in an elastic medium, $(\sigma_{ii}^b)_0 = -3K\Delta\alpha\Delta T$ as above, but the medium is assumed to be absolutely rigid,

$$i = r\varphi_0\,\Theta;$$

$\Delta\alpha = \alpha_b - \alpha_0$ is the difference between the coefficients of thermal expansion of the inclusion and the surrounding material, K is the bulk compression modulus, and G is the shear modulus.

Further approximate calculations were made of the thermal anisotropy stresses in hexagonal polycrystals for several special cases.

For a bicrystal with the hexagonal axes perpendicular and lying in the plane of the interface, the stresses are found from the following relations:

$$\sigma_\perp^{(1)} = \sigma_\perp^{(2)} = -\sigma_\parallel^{(1)} = -\sigma_\parallel^{(2)} = b_{\parallel,\perp}^{1,2}\,\Delta T, \tag{11}$$

where

$$b_{\parallel,\perp}^{1,2} = \frac{\alpha_\parallel - \alpha_\perp}{S_{11} + S_{33} - 2S_{13}}\,;$$

S_{11}, S_{13}, S_{33} are the elasticity coefficients in the corresponding directions of the hexagonal lattice, $\sigma_\parallel^{(1)}$ and $\alpha_\parallel$ are the stress and coefficient of thermal expansion along the hexagonal axis, $\sigma_\perp^{(1)}$ and $\alpha_\perp$ as above, but in a direction perpendicular to the axis. The superscript i = 1 is for the crystal on one side of the interface, while the index i = 2 is for the other side.

The values of $b_{\parallel,\perp}^{1,2}$ for zinc and cadmium are given in Table 8.

If the bicrystal has the hexagonal axes perpendicular and the interfacial planes coincident with the basal plane of one of the halves of the bicrystal, the stresses are found from the following relations:

$$\sigma_{\perp,\parallel}^{(2)} = -\sigma_{\perp,\parallel}^{(1)} = b_\parallel\,\Delta T, \tag{12}$$

$$\sigma_{\perp,\perp}^{(1)} = -\sigma_{\perp,\perp}^{(2)} = b_\perp\,\Delta T. \tag{13}$$

The numerical values of $b_\parallel$ and $b_\perp$ for zinc and cadmium are given in Table 8.

If we set $S_{12} = S_{13} = 0$, we have

$$b_\parallel = b_{\parallel,\perp}^{1,2} = a_0 = \frac{\alpha_\parallel - \alpha_\perp}{S_{11} + S_{33}},$$

$$b_\perp = 0.$$

Since $S_{11} = 1/E_{max}$ and $S_{33} = 1/E_{min}$, and replacing $\alpha_\parallel$ by α_{max} and $\alpha_\perp$ by α_{min}, we obtain

$$a_0 = \frac{\alpha_{max} - \alpha_{max}}{\dfrac{1}{E_{max}} + \dfrac{1}{E_{min}}}, \tag{14}$$

where α_{max} and α_{min} are the corresponding coefficients of linear expansion, and E_{max} and E_{min} are the corresponding moduli of elasticity of the first kind.

Table 8 gives the values of a_0 for zinc and cadmium, while Table 9 gives the values for several other metals.

The coefficient a_0 corresponds with the coefficient in the formula previously found by Boas and Honeycombe [57].

For the volume problem, where an anisotropic hexagonal prism (grain) is embedded in an isotropic matrix, V. A. Likhachev found the following relation:

$$\sigma_\perp = a_\perp \Delta T, \quad \sigma_\parallel = a_\parallel \Delta T, \tag{15}$$

where $\sigma_\perp$ and $\sigma_\parallel$ are the stresses perpendicular and parallel to the hexagonal axis,

$$a_\perp = \frac{1}{9\left(1 + \dfrac{3}{4}\dfrac{K}{G}\right)} \frac{(2S_{13} + S_{33})(\alpha_\parallel - \alpha_\perp)}{(S_{11} + S_{12})S_{33} - 2S_{13}^2};$$

$$a_\parallel = \frac{2}{9\left(1 + \dfrac{3}{4}\dfrac{K}{G}\right)} \frac{(S_{11} + S_{12} + S_{13})(\alpha_\parallel - \alpha_\perp)}{(S_{11} + S_{12})S_{33} - 2S_{13}^2}.$$

The values of $a_\perp$ and $a_\parallel$ for zinc and cadmium are given in Table 8.

As has already been pointed out, this calculation only gives approximate values for the stresses and the error may be a factor of 1.5 or 2.

Nevertheless, using these equations makes it possible to compare the thermostructural stresses occurring in different materials, and thus make an approximate calculation of the ability of a material to resist thermal fatigue.

It has been pointed out by V. A. Likhachev [58] that heating to 100-200°C produces stresses in excess of the flow limit, which shows that it is possible to have thermal fatigue from thermostructural stresses even for comparatively small temperature ranges.

Table 9 gives the values of a_0 for the special case mentioned previously. The highest values occur for uranium and the lowest for magnesium. The table also gives the ratio of the thermal anisotropy stresses $a_0^i \Delta T$ of the material in question to the same value for uranium for an equal temperature range, i.e., a_0^i / a_α^U. For example, under these conditions, the stress in zinc is a factor of 2 less, while in magnesium it is a factor of 127 less than in uranium.

Bearing in mind the difference in allowable operating temperatures for different materials, Table 9 gives the practically realizable (for heating above 0°C) temperature ranges, the stresses produced $a_0^i \Delta T^i$, and the ratio of these stresses to the corresponding value for uranium $a_0^i \Delta T^i / a_0^U \Delta T^U$.

TABLE 9. Thermostructural Stress Parameters for Various Metals

Metal	a_0, kg/mm²·deg	$\dfrac{a_0^i}{a_0^v}$	ΔT^i, °C	$a_0\Delta T^i$, kg/mm²	$\dfrac{a_0^i\Delta T^i}{v_0^v\Delta T^v}$
Uranium	0.254	1.0	550	139.7	1
Selenium	0.154	0.628	200	30.8	0.228
Zinc	0.125	0.506	400	50.0	0.375
Cadmium	0.0645	0.273	200	12.9	0.101
Tin	0.0506	0.206	200	10.1	0.0748
Tellurium	0.0322	0.131	400	12.9	0.101
Antimony	0.0150	0.0613	600	9.0	0.0733
Bismuth	0.0066	0.0269	250	1.65	0.0122
Magnesium	0.00193	0.00786	550	1.06	0.00786

For example, in uranium it is possible to have stresses up to 137 kg/mm², in zinc up to 50 kg/mm², in cadmium and tellurium up to 12.9 kg/mm², in tin up to 10 kg/mm², and in magnesium up to 1.06 kg/mm².

It is known to be a fact [59] that there is a large change in dimensions, and that failure cracks occur in cyclic heating of uranium, which shows that the thermostructural stresses reach large values in thermal fatigue.

Note that the formation of fatigue cracks is not alone determined by the magnitude of the thermostructural stresses, which depend primarily on the temperature range, but also by the absolute temperature level, since the mechanical properties of the crystals change with change in temperature.

Plastic deformation also exerts a substantial effect on the magnitude of the thermostructural stresses and on the nature of the stressed state in the various crystallites, which is not taken into consideration in the approximate analysis given above.

If a temperature gradient occurs across the body of the grains, the thermostructural stresses will combine with the stresses of the first kind in determining the total stress field. The solution of the problem of the combined effect of both types of stresses in thermal fatigue is quite complicated. There are as yet no quantitative methods for finding the overall effect of the stresses.

Comparison of Thermal and Mechanical Fatigue

Thermal and mechanical fatigue have much in common.

There are some investigators who try to extend the theory of mechanical fatigue to thermal fatigue. Thus, Coffin [14], in pointing out the importance of dislocations in thermal fatigue, assumes that failure occurs along the planes where the dislocations move back and forth. Clauss and Freeman [21] assume that thermal fatigue occurs in two stages: first, hardening of the material as a result of plastic deformation or microstructural dispersion hardening, and then weakening as a result of destruction of atomic bonds after the material has become quite hard.

However, there are important differences between thermal and mechanical fatigue:

1. Mechanical fatigue often occurs under a constant fixed load (with the nominal stresses fixed), while in thermal fatigue it is, as a rule, the deformations that are fixed, as determined by the temperature drop.

2. In thermal fatigue, the plastic deformation tends to be localized in the hottest spots, since nonuniform heating is occurring practically everywhere.

3. Local buckling may occur in thermal fatigue. Thus, for example, when the temperature is changed in a clamped sample, the largest amount of plastic compression deformation is concentrated in the hot part of the sample, where the flow limit is lowest. The plastic deformation increases the cross-sectional area. On cooling, the stress in this region is somewhat less than in the other parts of the sample, because of the larger cross-sectional area. Since the flow limit increases as the temperature drops, it is more difficult to produce plastic deformation in tension in this region than in the more stressed parts. Accordingly, the plastic tensional deformation is concentrated in the parts with lower cross-sectional area. This process is repeated cyclically, with the result that expansion

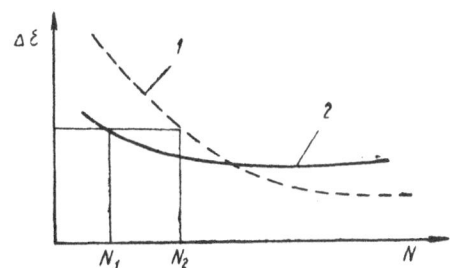

Fig. 48. Comparison of the fatigue curves of two materials: (1) first material; (2) second material.

occurs in the hotter part, and contraction everywhere else. This has been shown in the tests made by Clauss and Freeman [21]. Concentrating the tensile deformations in one region and the compressive deformations in others changes the fatigue characteristics of the material.

4. Cyclic temperature change in itself can have a considerable effect on the behavior of a material. An effect is to be expected in noncubic materials (zinc, uranium), but it has also been observed recently [60] in ordinary construction materials (steel, for example). It has been found that there is some temperature range in which cyclic temperature change can give different results from those found at a constant intermediate temperature.

5. In thermal fatigue, the deformation occurs under continuously changing temperature, which can have an effect on the structural processes and mechanical properties.

6. Mechanical and thermal fatigue tests are generally made at greatly different rates. Since a number of effects are enhanced by high temperature, they will be greatly affected by the duration of the cycle.

Accordingly, the results of thermal fatigue tests may be partially interpreted from the results of mechanical fatigue tests made with a small number of cycles. However, complete correlation cannot exist because of the differences pointed out above in the conditions under which the tests are made. Because of the possibility of concentrating the deformations and of localizing the tensile and compressive deformations, the duration in thermal fatigue may be less than in mechanical fatigue for the same mean plastic deformation per cycle.

Since thermal fatigue usually occurs at a relatively small number of cycles (10^3-10^4 cycles), it must be regarded as a process involving large plastic deformations, which corresponds with the left branch of the fatigue curve. However, in mechanical fatigue, the operating life of the part corresponds with a considerably larger number of cycles. This means that it is necessary to find the true fatigue limit or maximum endurance at a large number of cycles, i.e., investigating the right-hand branch of the fatigue curve.

It is thus often impossible to predict the behavior of a material under thermal fatigue from mechanical fatigue tests expressed in terms of endurance limits. The fatigue curves of two materials, given in Fig. 48, show that the endurance calculation gives different results at small deformations from those at large deformations.

Methods of Making Thermal Fatigue Tests

The methods that have so far been developed for thermal fatigue tests may be divided into three basic groups:

1. Qualitative methods of finding the thermal fatigue resistance of materials.

2. Natural methods of thermal fatigue testing, which reproduce the operating conditions.

3. Quantitative methods of finding the thermal fatigue resistance of materials by determining the deformations and stresses.

Qualitative Methods

The earliest thermal fatigue tests were made to get a qualitative value for the resistance to cyclic temperature change, using samples of the simplest possible shape. These methods, which differed in the method of heating and cooling and in the form of the samples, did not contemplate anything more than finding the number of cycles required to produce cracks, but in some cases a determination was made of the change in shape and dimensions of the samples as a function of the number of cycles. The magnitude of the thermal stresses and deformations occurring in these tests was determined by the combined action of all the factors mentioned above and practically defied analysis [4, 6, 12, 24, 40, 59, 61]. V. I. Zalesskii and D. M. Korneev [4] investigated the heat resistance of 7KhZ, 4KhNV, 35KhGSA, and 4KhVS steels. Cylindrical samples of these steels, 30 mm in diameter, were heated to 800°C in a lead bath, and cooled in running water. In addition to flaw formation, there was a change in dimensions and shape of the samples visible to the naked eye. It was also noted that when cylindrical samples were heated along the edge above the critical point AC_1, the cracks from structural and thermal stresses appeared earlier than when there were no structural transformations (heating below AC_1).

D. I. Kostenko [6] made a study of the behavior of 5KhNT and 5KhNM steels under variable temperatures. The samples used were 25 mm in diameter and 100 mm long. The samples were heated with high-frequency current and cooled with water. The tests showed that when the hardness is high, these steels have high sensitivity to formation of thermal fatigue cracks.

It was also found that in dense steel samples the cracks occurred later and developed more slowly than in samples of less dense steel. Cracks also formed and developed more slowly if the samples had a clean surface.

M. V. Pridantsev and A. R. Krylova [40] built a merry-go-round type of outfit for making thermal fatigue tests on thin sheet materials. Samples $120 \times 120 \times 1.5$ mm with five holes in the center (to make the tests more "severe") were heated to a known temperature on a gas burner for 1 minute, after which the sample holder was moved to the next position, where the samples were cooled with air for 1 minute. The tests were made on a type Kh20N80 alloy. It was found that the thermal resistance of the alloy decreased with increase in thickness of the sheet. Testing by heating with a gas burner is widely used, since it is an easy and simple method of reaching high temperatures. However, tests with this type of heating usually show a large spread in the results because of inaccuracy in measuring the temperature, difficulty in automating the test procedure, etc.

M. Ya. L'vovskii and I. A. Smiyan [61] have developed a method for finding the resistance of sheet materials to thermal cycling. The sample, consisting of a narrow plate with a 0.1-mm radius slot in one end, was subjected to repeated heating and cooling. The heating was done in an electric furnace, and the cooling was done with water. The alloy tested was ÉI435, 1.5 mm thick.

In the vast majority of cases, the results of these qualitative test methods cannot be used to predict the behavior of structural elements of any given material under actual operating conditions, since changing the dimensions and shape or the heat-transfer conditions can produce a substantial change in the stressed state at the danger zones.

Usually, the results of the tests, when made under definite conditions, are used to make a comparison between the thermal fatigue resistance of different materials. However, changing the test procedure may have the result that the material found to be the best when tested in one way turns out to be less good when tested some other way [1].

Natural Tests

Natural tests may be made either under operating conditions, or under laboratory conditions which mock up the operating conditions. These tests serve to show the behavior of parts made of a definite material under definite operating conditions. The tests are undoubtedly useful for mass-produced parts, or for very important structural elements. The results of the experiments give a direct indication of the strength of the structure under definite operating conditions. However, it is very difficult to use the results for other operating conditions, since the quantity measured (number of cycles to failure) reflects the combined effect of different factors, such as the material, the way the heat is propagated, the method of construction, etc.

Thus, Weisberg and Soldan [62] made thermal fatigue tests on welded pipe joints in ferrite and austenite steels right in the steam pipes of a power plant.

Bentele and Lowthian [63] made a thermal fatigue study on valve heads cast from austenite steel. The valve heads were heated with gas to 800-900°C for 90 sec, and then cooled with air for 30 sec to 20°C. Fatigue cracks appeared in the valves after 500 cycles when tested in this way.

I. A. Oding and Yu. V. Kostochkin [64] have developed a method and apparatus for testing turbine blading and samples under alternating temperature in a gas stream. Four blades may be tested at once in the apparatus, where they alternately have hot gas blown over them from a heating chamber, and cold air taken from the atmosphere.

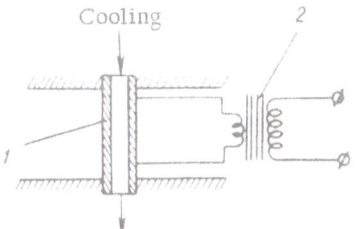

Fig. 49. Clamped tubular sample for thermal fatigue tests [14]: (1) samples; (2) transformer.

Tests were made on blades of ÉI765 alloy, which were heated to 750°C and cooled to 70°C. After a definite number of temperature cycles, the blades were mounted on the rotor of an experimental gas turbine setup and run to failure.

Quantitative Tests

A number of tests have been made, including an analysis of the stressed and deformed states, with the stresses and deformations during thermal fatigue determined analytically or experimentally. These tests make it possible to get a quantitative picture of the behavior of the material, which gets close to solving the problem of designing parts for thermal fatigue from the results of tests made on samples.

Figure 49 shows a schematic diagram of Coffin's setup [14]. The tubular sample is rigidly clamped at the ends and is subjected to cyclic heating and cooling. The heating is done by passing current through the sample. For this purpose, a transformer was used which gave a current of 1000 A at a potential of 2 V. The cooling was done by passing air through the hole in the sample. The sample was clamped longitudinally by means of a rigid system consisting of two compression flanges held together by a pair of columns. The rigidity of the system is so great that the total mechanical deformation of the sample was practically equal to the thermal deformation:

$$\varepsilon = \varepsilon_{el} + \varepsilon_{pl} \simeq \varepsilon_{T} = \alpha \Delta T, \tag{16}$$

where ε_{el} is the elastic deformation of the sample, ε_{pl} is the plastic deformation, $\varepsilon_{T} = \alpha \Delta T$ is the thermal deformation, α is the coefficient of thermal expansion, and ΔT is the temperature difference.

The stresses in the sample were found from tensometers, fastened to the surface of the columns. Knowing the magnitude of the stresses, the elastic deformation could be calculated, assuming E = const, and, subtracting it from the total deformation gave the plastic deformation.

This outfit was used to make tests on 347 stainless steel for different temperature differences, the mean temperature T_m remaining constant. The number of cycles to failure was found as a function of the temperature difference and the corresponding deformation.

The outfit was built in such a way that different parts of the thermal cycle could be set for different times, from 1 to 60 sec.

It was also intended that the outfit could be used to make mechanical fatigue tests at a constant elevated temperature of T = 350°C. The mechanical loading of the sample was accomplished by cyclic heating and cooling of the columns holding the compression flanges together.

Particular attention was given to the shape and dimensions of the sample. The working part of the sample was a thin-walled cylinder in which analysis showed the stresses occurring from radial heat transmission during heating and cooling to be about 0.5 kg/mm². In order to reduce stress concentration, the transition from the working part to the heads was made in the form of a smooth curve. The axial temperature gradient produced stresses in the heads of about 0.7 kg/mm². These stresses were small in comparison with the stresses produced by restraining the deformation of the sample in a longitudinal direction and, thus, what occurred in the sample was practically a uniform uniaxially stressed state. This made it possible to find the behavior of a volume element of the material under thermal fatigue. Account was also taken of the possibility of loss of strength in the sample during the high-temperature part of the cycle. It was found that longitudinal bending occurred for a deformation of several percent, while in Coffin's experiments the deformation did not exceed 1%.

The results of Coffin's experiments are widely discussed in various sections of this book. In spite of their great worth, they only represent the special case where the thermal deformation produces an equal mechanical deformation when the sample is clamped. Because of this, Coffin was obliged to compare the behavior of a material for different amplitudes of deformation with different temperature differences, which does not make it

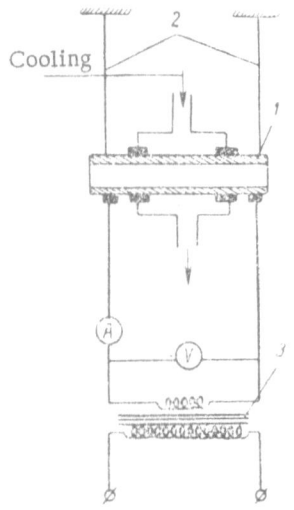

Fig. 50. Diagram of thermal fatigue tests on tubular samples without clamping [16]: (1) sample; (2) flexible supports; (3) transformer.

possible to find the effect of each factor (magnitude of deformation and temperature difference) separately.

The paper by V. N. Kuznetsov [16] takes as its problem finding experimental ways of justifying the tolerances on cyclic temperature stresses in the often encountered case of radial heat flux in a tube. The thermal fatigue tests were made on the outfit shown schematically in Fig. 50.

The tubular sample 1, 6 mm in diameter with wall thickness 1 mm, made of 1Kh18N9T austenite steel was suspended on the flexible supports 2. Heating the sample with electric current while continuously cooling the outside surface produced a temperature drop of up to 550°C between the inner and outer surfaces. Turning the transformer 3 on and off periodically, loaded the wall of the sample with periodic thermal stresses.

During the test, the sample could expand freely along its axis. Accordingly, during cooling, the outer surface layers of the sample were in a plane stressed state. The mechanical deformation of the metal on the outer surface of the tube was found from the conditions

$$\int_{R_1}^{R_2} \sigma r \, dr = 0, \tag{17}$$

$$\varepsilon + \int_0^t \alpha \, dt = \varepsilon_z = \text{const}, \tag{18}$$

[here, the experimental function $\sigma = f(\varepsilon)$ was known for t = const], where t is the temperature in the wall of the tube, r is the coordinate along the radius of the tube, σ is the stress in the wall of the tube, ε is the mechanical deformation, R_1 and R_2 are the inner and outer radii of the tube, respectively, $\int_0^t \alpha \, dt$ is the thermal deformation of the surface layer which would occur if the surface layer could elongate freely as the result of heating to the temperature used, and ε_z is the axial deformation resulting from axial thermal stresses.

In spite of the value of this method, which simulates the actual conditions under which pipes operate in heat exchange apparatus, the difficulties must be pointed out which arise in calculating the stresses and deformations under cyclic temperature changes.

Cracks appeared earliest on the outer surface of the sample, where the largest deformations were acting. Periodic observation of the sample made it possible to establish the instant at which the cracks started to form, as well as the instant at which a well-developed network of cracks occurred.

In the paper by A. V. Ratner [51], an attempt was made to find a criterion for the failure of tubular samples of various thicknesses made of carbon steel and 1Kh18N9T and 15KhM steels under cyclic temperature change.

The heat load (q = 100,000 kcal/m²-h) was produced by continuously passing steam inside the samples at a pressure of 100 atm and a temperature of 480-500°C. The outer surface of the pipes was cooled periodically with chemically pure water. The duration of the cycle was 1 min 54 sec. It was observed in testing samples of different thicknesses under the same conditions of heating and cooling that surface failure occurs at the same number of cycles if the deformation reaches a definite critical value ε_{cr}.

For austenite steel át 5400 cycles, ε_{cr} was 0.43% on the inner wall of the pipe, which is a factor of 3.5 greater than the elastic deformation. For carbon steel at 8826 cycles, $\varepsilon_{cr} = 0.21\%$, which is a factor of 2 greater than the elastic deformation. On the outer walls cooled with water, ε_{cr} was less than the above values for both austenite and carbon steel.

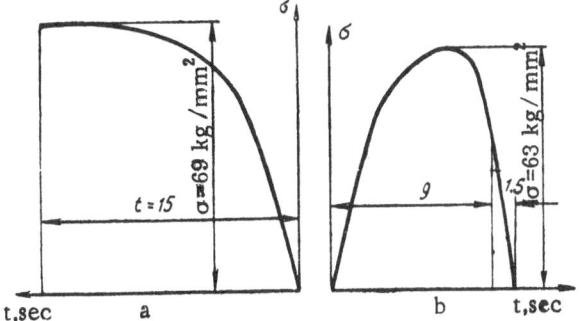

Fig. 51. Oscillograms of stress change during a thermal cycle [15]: (a) tension on cooling (from 750 to 100°C); (b) compression on heating (from 100 to 750°C).

The deformations were calculated from the nominal elastic stresses, which gives only an approximate value. These calculations do not provide any basis for a criterion of failure.

A further development of Coffin's method was to set up an outfit on which tests could be made under less rigid conditions than those existing when the sample is tested with the ends clamped [15]. The difference between this outfit and that of Coffin, where the ends of the sample were rigidly clamped, is that elastic membranes are connected in series with the sample, thus making it possible to regulate the rigidity of the loading. In the tests on this outfit, deformation of the sample is equal to

$$\varepsilon \approx \alpha \Delta T - kP, \tag{19}$$

where α is the coefficient of linear expansion, ΔT is the temperature drop, k is the coefficient of rigidity of the system, and P is the force on the sample during deformation.

Changing the rigidity of the system makes it possible to get deformations of different amplitudes in the sample for the same temperature cycle. Thus, it is possible to construct a fatigue curve expressed in terms of the amplitudes of the elastoplastic deformations occurring in a definite range of temperature change.

Note that it is not possible to get mechanical deformations greater than $\varepsilon = \alpha \Delta T$ on this outfit. Considering the results of Coffin's experiments [14], which show that there is a change in the stresses during a thermal fatigue test, it is to be expected that the force P on the sample will change, which produces a change in the magnitude of the elastoplastic deformation occurring in a given test scheme.

The outfit gave deformation diagrams for tension and compression under varying temperature. The following functions were needed to get the diagrams:

Stress change with time, $\sigma = f_1(t)$,
Temperature change with time $T = f_2(t)$,
Change in the coefficient of thermal expansion with time, $\alpha = f(t)$.

These functions may be used to find the deformation, when the deformation occurs rigidly

$$\varepsilon = \alpha T = \varphi(t) \tag{20}$$

and the deformation when the rigidity is limited

$$\varepsilon = \varphi(t) - m(t). \tag{21}$$

By using the functions, the curve $\sigma = F(\varepsilon)$ may be found.

S. V. Serensen and P. I. Kotov [15] used this method to obtain deformation diagrams for ÉI437B alloy with rigid loading, and the temperature changing from 100 to 750°C. For this alloy

$$\alpha = 12 \cdot 10^{-6} + 1.27 \cdot 10^{-8} \, T, \tag{22}$$

$$\varepsilon = 12 \cdot 10^{-6} \, T + 1.27 \cdot 10^{-8} \, T^2, \tag{23}$$

Curves of the stress changes corresponding with the equation $\sigma = f_1(t)$ were taken with an oscillograph (Fig. 51). The temperature cycle [curves $T = f_2(t)$] was taken with a recording instrument. These curves and the equation for ε give the $\varepsilon = \varphi(t)$ curves (Fig. 52). Then deformation diagrams were constructed for tension and compression.

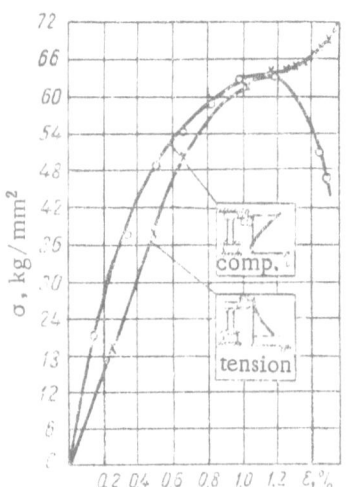

Fig. 52. Deformation diagrams for ÉI437B alloy under temperature change [15].

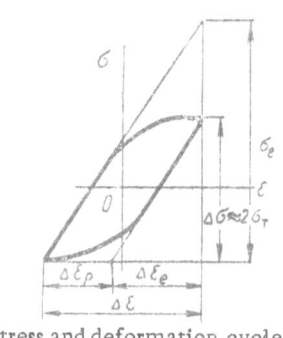

Stress and deformation cycle

Fig. 53. Example of a thermal fatigue calculation.

In making the experiments which give the fatigue curve in the form "plastic deformation vs. number of cycles to failure" for the same temperature range and temperature level, Coffin [23] set up an outfit in which the temperature cycle is accompanied by a mechanical cycle produced by a special eccentric mechanism. The mechanism is guided by the electrical circuit used to control the heating and cooling so that the heating and cooling can be combined with the mechanical tension − compression cycle to give a definite program. Various types of cycles could be produced on the outfit, for example a cycle in which the temperature is raised at the instant tension occurs, as opposed to what was done in the earlier method.

In [65], a description is given of a machine that simulates thermal fatigue under conditions of pure displacement by means of cyclic heating of the sample, synchronized with cyclic mechanical loading.

The amount of mechanical deformation may be varied independently of the temperature cycle. The machine may be used to measure the stresses during deformation. This makes it possible to construct fatigue curves plotted as "deformation − number of cycles" or "stress − number of cycles."

Making tests approximating the actual conditions where a temperature gradient is produced over the cross section means developing accurate methods of calculating the stresses and deformations in the elastoplastic range with nonuniform cyclic change of the deformation field under changing temperatures.

Thermal Fatigue Calculations

No reliable method has yet been worked out for designing for thermal fatigue. As an approximate estimate of the behavior of a part under cyclic temperatures, we give the method of calculation suggested by Coffin [66].

It is proposed for this purpose to use an experimental relation of the form

$$N^{1/2} \Delta \varepsilon_p = C, \qquad (24)$$

where N is the number of cycles to failure, $\Delta \varepsilon_p$ is the change in plastic deformation per cycle, and C is a constant, depending on the material.

In order to design the part, we need to know $\Delta \varepsilon_p$ at the most "dangerous" point.

The calculation of $\Delta \varepsilon_p$ was made in the following order:

1. A calculation was made of the maximum elastic stress σ_e, occurring for the total temperature change from T_{min} to T_{max}, under the assumption that the material is perfectly elastic, even if the stress calculated turned out to be greater than the flow limit of the material.

2. The elastic thermal stresses were converted to deformation by dividing by the modulus of elasticity. The deformation was assumed to be equal to the total change in deformation, $\Delta \varepsilon$.

3. A calculation was made of the elastic fraction $\Delta \varepsilon_e$, of the deformation change, under the assumption that the stresses in the material under tension or compression cannot exceed the flow limit for the mean temperature. Thus, the value of $\Delta \varepsilon$ was found by dividing twice the flow limit by the modulus of elasticity.

4. The plastic deformation change $\Delta \varepsilon_p$ was found by subtracting $\Delta \varepsilon_e$ from $\Delta \varepsilon$.

5. The number of cycles to failure for the material was calculated from the formula

$$N = \frac{C^2}{(\Delta \varepsilon_p)^2}. \qquad (25)$$

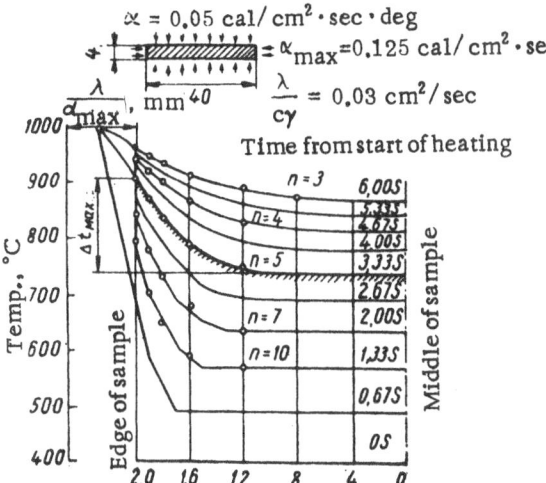

Fig. 54. Temperature distribution as a function of time in a sample of rectangular cross section which is rapidly heated from an initial temperature of 400°C by bringing the surface in contact with a gas at a temperature of 1000°C [67]: (α) heat transfer coefficient; (λ) coefficient of thermal conductivity; (γ) density; (c) specific heat; (n) exponent of the parabola.

The allowable number of cycles [N] was found using an appropriate safety factor after making an approximate calculation of the deformations as well as allowing for the possibility of change in operating conditions and nonuniformity of the material.

Coffin points out that the above order has been used in designing various structural elements such as gas turbine blading and nuclear reactor parts, where large thermal stresses occur.

Consider the following example of a calculation (Fig. 53).

The exhaust temperature of a diesel engine varies from 375 to 650°C. Analysis of the stresses has shown that in a steel containing 13% chromium the thermal stresses reach 84 kg/mm^2.

The elastic thermal stress is $\sigma_s = 84$ kg/mm^2 (elastic analysis).

The total deformation change is

$$\Delta\varepsilon = 84/[(2.1) \times 10^4] = 0.004.$$

Assume that the flow does not exceed the elastic deformation change

$$\Delta\varepsilon_e = (2 \times 21)/[(2.1) \times 10^4] = 0.002.$$

The plastic deformation change is

$$\Delta\varepsilon_p = \Delta\varepsilon - \Delta\varepsilon_e = 0.02.$$

The number of cycles to failure is

$$N^{1/2} \Delta\varepsilon_p = C = 0.57,$$

$$N^{1/2} = 0.57 \div 0.002 = 285,$$

$$N = 81,000.$$

The paper by Haas [67] gives a method of finding the maximum temperature drop for which thermal fatigue failure does not occur. The calculation is made for a plate of rectangular cross section, subjected to cyclic heating by a hot gas.

Figure 54 gives the time dependence of the temperature in the plate for an initial temperature of 400°C and a gas temperature of 1000°C. The temperature difference between the inner region and the surface of the sample, and the temperature gradient at the surface decrease with time.

The curves in Fig. 54 are replaced by nth-order parabolas (for simplicity in integration), which makes it a simple matter to find the stresses.

On the surface of the sample, the stresses are equal to

$$\sigma_R = - E_t \beta_t \Delta t_{max} \frac{n}{n+1}. \tag{26}$$

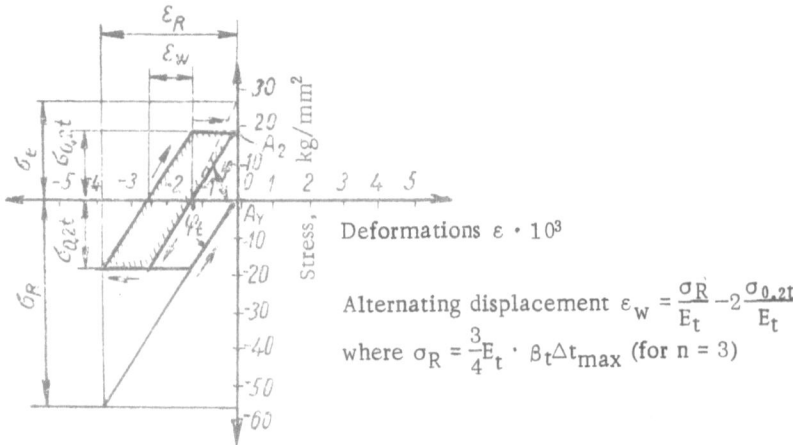

Alternating displacement $\varepsilon_w = \dfrac{\sigma_R}{E_t} - 2\dfrac{\sigma_{0.2t}}{E_t}$

where $\sigma_R = \dfrac{3}{4}E_t \cdot \beta_t \Delta t_{max}$ (for $n = 3$)

Fig. 55. Alternating displacements in the outer fibers of a sample under rapid heating and slow cooling: (σ_R) calculated stress in the outer fibers; (ε_R) total deformation in the outer fibers; ($\sigma_{0.2t}$) provisional flow limit at the temperature t; (σ_e) internal stress at room temperature t; (E) modulus of elasticity at temperature t; (β_t) coefficient of linear expansion at temperature t; $\tan\varphi = E$; $\tan\varphi_t = E_t$; (Δt_{max}) greatest temperature difference; (A_1) starting point of the first deformation process; (A_2) starting point of the second and each subsequent deformation process [67].

Figure 55 shows a diagram of the deformation during rapid heating and subsequent gradual equalization of temperatures under the assumption that the material is ideally plastic. Here the maximum deformation corresponds with the existing (elastic) stress σ_R. Thus, for each temperature change, the external fibers take up the amount of deformational work given by the area of the cross-hatched part of the diagram in Fig. 55.

The segment ε_w corresponds with repeated plastic deformation (alternating slip), and hence the case under discussion refers to the left-hand side of the fatigue curve. Therefore, an important factor is to have high plasticity in the material and no temperature range showing clearly defined brittleness.

The importance must also be noted of high corrosion resistance under the action of hot gases (especially the danger of intercrystalline corrosion). Under conditions where corrosion, recrystallization, and other structural changes are possible, alternating plastic deformations must be eliminated to avoid flaw formation.

This is expressed by the condition

$$|\sigma_R| \leqslant 2\sigma_{0.2t}. \tag{27}$$

From Eqs. (26) and (27) we obtain the maximum allowable temperature drop

$$\Delta t_{max} = \frac{n+1}{n}\,\frac{2\sigma_{0.2t}}{\beta_t E_t}, \tag{28}$$

where t is the temperature of the surface.

If rapid heating is accompanied by rapid cooling, the allowable temperature drop is considerably less.

Table 10 gives results of calculating Δt_{max} for different materials where the temperatures depend on the time (as in Fig. 54). It is easily seen that as the temperature of the surface is raised, Δt_{max} decreases.

Material No. 4 (Nimonic 80A), which has a high value of $\sigma_{0.2t}$, small β_t, and not very high E_t, shows the largest Δt_{max}.

TABLE 10. Maximum Temperature Drop for Various Materials [67]

Material	Chemical composition of material, %								Edge temp., t = 700°C				Edge temp., t = 800°C				Edge temp., t = 900°C			
	C	Cr	Ni	Co	Mo	V	Ti	Others	Modulus of elasticity* $\times 10^3$ kg/mm^2	Coeff. of linear expansion $\times 10^{-6}$ 1/°C	Flow limit* $\sigma_{0.2}$, kg/mm^2	Temperature drop, Δt_{max}, °C	Modulus of elasticity* $\times 10^3$ kg/mm^2	Coeff. of linear expansion $\times 10^{-6}$ 1/°C	Flow limit* $\sigma_{0.2}$, kg/mm^2	Temperature drop, Δt_{max}, °C	Modulus of elasticity* $\times 10^3$ kg/mm^2	Coeff. of linear expansion $\times 10^{-6}$ 1/°C	Flow limit* $\sigma_{0.2}$, kg/mm^2	Temperature drop, Δt_{max}, °C
1**	0.1	18	8	—	—	—	—	—	14	18.5	12	123	13	19	10	108	11.5	19.5	7	83
2**	0.15	25	20	—	—	—	—	—	14.8	17.0	16	170	14	17.5	13	141	13.0	17.8	9	104
3**	0.5	18	9	—	—	1.0	—	—	13.5	18.5	24	256	12.5	19	15	169	11.5	19.5	12	143
4***	0.1	20	Re-mainder	—	—	—	2.5		15.5	14.5	53	627	11.5	15	41	634	9.9	15.8	19	323
5***	0.28	20	10	45	2.2	3.0	—	1.4	—	—	—	—	16.5	16	44	445	—	—	—	—

* At the temperature of the edge.
** Tempering.
*** After dispersion hardening.

101

Material	Mean operating temp., °C	Max. temperature difference, °C
Beryllium	20	70
Magnesium	20	100
Aluminum	20	9
Titanium	20	780
Titanium	400	350
Vanadium	20	690
Cobalt	20	105
Nickel	20	60
Nickel	600	60
Copper	20	30
Zirconium	20	260
Molybdenum	20	500
Molybdenum	600	380
Beryllium oxide	600	185
Stainless steel	20	150
Stainless steel	400	75

Under conditions where deformation is completely prevented, the maximum change in the temperature difference, calculated on the assumption that the material will withstand an unlimited number of temperature cycles, is given by the following relation [68]:

$$\Delta T = \frac{2\sigma_y}{\alpha E},\tag{29}$$

where E is the modulus of elasticity of the first kind, α is the coefficient of linear expansion, and σ_y is the elastic limit.

Some values of ΔT for various materials are given in Table 11.

Coffin [14] gives a method of calculating the number of cycles to failure in thermal fatigue, under conditions where the stresses and deformations are concentrated, for plates with a circular opening, where the stress and deformation concentration coefficient in the elastic region is equal to three. It is assumed that the deformation concentration coefficient does not change in the plastic range.

Using the equation

$$N^{0.5}\Delta\varepsilon_p = 0.359.\tag{30}$$

Coffin makes a comparison between the conditions at the edge of the opening and at points far from the opening:

$$N_1^{0.5}\Delta\varepsilon_{p1} = N_2^{0.5}\Delta\varepsilon_{p2}\tag{31}$$

where N is the number of cycles, $\Delta\varepsilon_p$ is the change in plastic deformation per half cycle, and 1 and 2 are points, respectively located far from and close to the deformation concentrator.

By rearranging the last expression, we can obtain:

$$N_1^{0.5}\left(\Delta\varepsilon_\tau - \frac{\Delta\sigma_1}{E}\right) = N_2^{0.5}\left(k\Delta\varepsilon_\tau - \frac{\Delta\sigma_2}{E}\right).\tag{32}$$

Here the plastic deformation change is replaced by the total deformation change less the elastic part. Further,

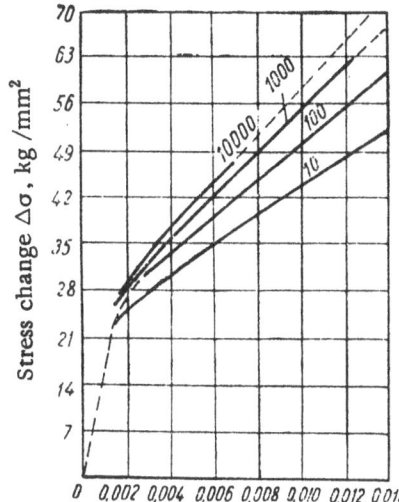

Fig. 56. Change in deformation as a function of stress for annealed 347 stainless steel under thermal cycle [14].

$$\Delta \varepsilon_{T1} = \alpha \Delta T, \tag{33}$$

where α is the coefficient of thermal expansion at the point 1, far from the stress concentrator, and $\Delta \varepsilon_{T_2} = k \alpha \Delta T$ is the total deformation change at the point 2 (at the point where the deformation is concentrated). Therefore,

$$\frac{N_1}{N_2} = \frac{Ek\alpha \Delta T - \Delta \sigma_2}{E\alpha \Delta T - \Delta \sigma_1}. \tag{34}$$

To solve Eq. (34), a particular value is taken for $\alpha \Delta T_0$. N_1 is found for this value in Fig. 14 (curve for sample with no stress concentrators). Figure 56 gives the stress change $\Delta \sigma_1$ for the total deformation change $\Delta \varepsilon_T = \alpha \Delta T_0$.

Further, for $\Delta \varepsilon_{T_2} = k \alpha \Delta T_0$ (where k = 3), Fig. 56 gives values of $\Delta \sigma_2$ for different values of N. Then for each value of $\Delta \sigma_2$ and the values of $\Delta \sigma_1$, N_1, k, and $\alpha \Delta T_0$, N_2 is found from Eq. (34).

The method of successive approximations may be used to find the special value of N_2 which satisfies Eq. (34) in Fig. 56. The calculated values are designated in Fig. 14 by the index "method A."

Predicting the Behavior of a Material and Ways of Increasing the Strength under Thermal Fatigue

The most complete and accurate information on the behavior of a material in actual service may be found by making a thermal fatigue experiment under conditions approximating those actually encountered.

However, in choosing a material to operate under cyclically varying temperatures, and hence under cyclically varying stresses and deformations, an approximate qualitative estimate may be made, based on an analysis of the ordinary mechanical characteristics.

An important property in this respect is the plasticity of the material. Increasing the plasticity as a rule increases the resistance to plastic deformation of alternating sign under cyclic temperature action.

It is also to be assumed that creep deformation will have a beneficial effect in removing the thermal stresses when the cycles are of long duration [69]. From this point of view it is useful to know something about the creep characteristics.

It has already been shown that predicting the thermal fatigue behavior of materials from the absolute endurance limits does not give correct results, since the number of thermal cycles is not greater than 10^5 in power plants. A comparison may be made from the limiting values of endurance, approximate values of which may be calculated from formulas holding for $N < 10^6$ [70]:

for tension

$$\sigma_{-1N} = \sigma_b - \left(\frac{\sigma_b - \frac{\sigma_{-1}}{k_f}}{6} \right) \log N; \tag{35}$$

for bending

$$\sigma_{-1N} = \frac{\sigma_b}{\left(\frac{N}{1000} \right)^{1/3 \log \frac{k_f \sigma_b}{\sigma_{-1}}}}, \tag{36}$$

where σ_b is the time resistance; σ_{-1} is the corresponding absolute endurance limit, k_f is the corresponding effective stress concentration factor, and N is the number of cycles from which the limiting value of endurance is determined.

With no stress concentrations, k_f may be set equal to unity.

The existing experimental and theoretical data enable constructional and engineering recommendations to be made as to increasing the strength under changing temperature.

As in the case of parts subject to mechanical loading alone, large concentrations of stresses and deformations must be avoided as far as possible.

In order to decrease thermal stresses and deformations, it is desirable for the part to be shaped in such a way that the heating and cooling is as uniform as possible, i.e., the temperature gradient is low over the cross section and along the length. It is also necessary to get the least amount of rigidity in the external mechanical couplings, which tend to prevent thermal deformation of the part under thermal action. In particular, in welded, riveted, and other joints, a necessary condition is to have smooth transitions where the parts are joined.

We now pass to a consideration of the importance of engineering factors.

As we know, the surface treatment has a substantial effect on the fatigue strength, and so, bearing in mind that in many cases the surface layers of a part are the ones that are most heavily loaded in thermal action, a high quality of surface treatment must be sought. If there are no experimental data on thermal fatigue, an approximate estimate of the behavior of the material may be made from the results of mechanical fatigue tests.

Since, in cyclic temperature action the stresses are set up by restraining the deformation, a favorable effect is exerted by cold working for relatively small deformations.

It has already been pointed out that some coatings have a favorable effect in thermal fatigue. Naturally, reducing the thermal stresses in parts increases the resistance to thermal fatigue.

Using a material which combines low modulus of elasticity, small coefficient of thermal expansion, and high thermal conductivity will greatly lower the thermal stresses. Titanium is an example of a material of this type.

STRENGTH UNDER CYCLIC TEMPERATURE ACTION WITH SIMULTANEOUS STATIC LOADING

In practical operation of machines and equipment, the structural elements are often under cyclically varying temperatures which produce thermal stresses, with a static load present at the same time.

Superimposing static stresses on a thermal stress cycle makes the cycle asymmetric, which means a special approach to calculating the strength under these conditions.

An example of this is provided by gas turbine blading which is under centrifugal force and thermal stresses at the same time. The thermal stresses may change rapidly as a result of heating and cooling when the turbine is started and stopped.

Further, a discussion will be given of strength calculations for the conditions existing in power generating equipment.

Strength Calculation

The design of housings of power generating equipment operating under internal pressure is usually made from formulas which do not contain the thermal stresses directly, the assumption being made that the operating stresses remain practically constant and that the effect of temperature changes is small.

In designing nuclear reactors, however, the question arises as to what thickness to take for the structural parts of fuel elements. Large thermal stresses occur if the parts are thick. It would seem that the permissible value of these stresses ought to be found from the usual formulas for vessels operating under pressure. But the possibility must be taken into account that the stresses will be lowered from plastic deformation. In view of this, the thermal stresses may practically be neglected under steady state conditions. However, if the reactor power is changed, plastic deformation of alternating sign occurs, and the strength must be calculated from the fatigue limit for a limited

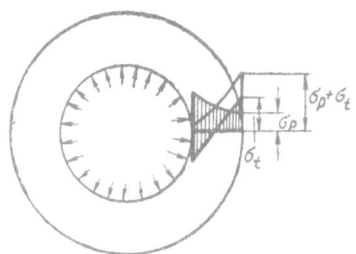

Fig. 57. Stresses in a ring.

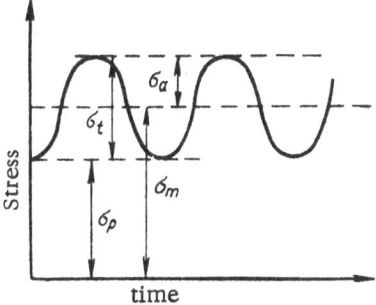

Fig. 58. Change in stresses.

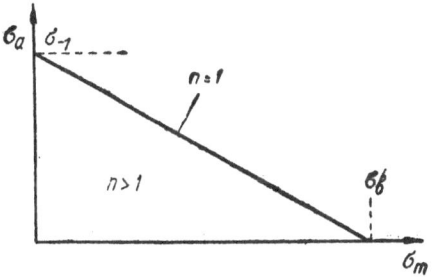

Fig. 59. Diagram showing limiting cycles.

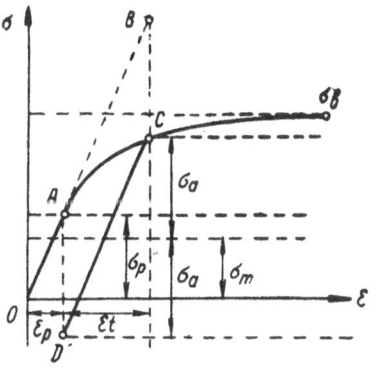

Fig. 60. Deformation diagram [70].

number of cycles, bearing in mind the practical possibility of repeated power changes (about 10^3 times) during the lifetime of the equipment.

A consideration of the conditions under which thermal stresses occur shows that these stresses may be substantially greater in nuclear reactor parts than in ordinary power generating and engineering equipment. Actually, in reactors, the intense radiation generates heat inside the material of the fuel elements and the surrounding components, while in ordinary equipment the heat is supplied from the outside and transferred to the surface. Further, the heat generation rate in a reactor may change very rapidly, while in ordinary heat power installations the increase in temperature level is limited by the rate at which fuel is added.

An analysis of the stressed state under the combined action of static load and cyclic thermal stresses is made on a thick-walled ring loaded with an internal pressure (Fig. 57). The pressure produces tangential stresses σ_p on the surface of the ring, just as in the walls of a thick-walled vessel. These stresses cannot exceed the flow limit.

In addition, let the ring be nonuniformly heated, with the temperature higher on the inner surface than on the outer. The result is to produce additional peripheral tensile stresses σ_t in the outer part of the ring, which add to the stresses σ_p. If the sum $\sigma_p + \sigma_t$ is greater than the flow limit, the actual stress on the outer surface cannot be calculated by simple addition.

Assume that under constant radial load the inner surface of the ring is at a higher temperature than the outer. If $\sigma_p + \sigma_t < \sigma_T$, the stresses on the outer surface will change as shown in Fig. 58. The characteristics of the stress cycle are calculated in the following way:

$$\text{mean stress} \quad \sigma_m = \sigma_p + \frac{\sigma_t}{2}, \tag{37}$$

$$\text{amplitude} \quad \sigma_a = \frac{\sigma_t}{2}. \tag{38}$$

The fatigue strength of the outer layers may be verified on the limiting cycle diagram (Fig. 59). The equation of the straight line giving the limiting cycles is written in the form:

$$\frac{\sigma_a}{\sigma_{-1}} + \frac{\sigma_m}{\sigma_T} = 1. \tag{39}$$

The safety factor is

$$n = \frac{1}{\dfrac{\sigma_a}{\sigma_{-1}} + \dfrac{\sigma_m}{\sigma_T}}. \tag{40}$$

We shall now consider the case where the maximum stress exceeds the flow limit.

Langer [70] has proposed the following approximate method of calculation using the deformation diagram under load in the plastic range (Fig. 60). The deformations ε_p and ε_t are found from the

105

calculated elastic stresses σ_p and σ_t. When the inner surface of the ring is heated, the deformations increase from ε_p to $\varepsilon_p + \varepsilon_t$, and the actual stresses are given by the curve OAC (instead of the straight line OAB). At the instant where there is no temperature drop, the deformations again decrease to the value ε_p and the stresses follow a linear law (the line CD). Further temperature changes produce cyclic stresses and deformations which follow the line DC (CD). Accordingly, in making the calculation, the mean stress σ_m must be reduced by the amount corresponding with the segment BC.

Consider now an example of a calculation using the above method.

In a thick-walled steel cylinder (347 steel) in which the outer layers are heated by induction, while the inner layers are cooled with water at a pressure of 180 atm, the calculated elastic stresses on the inner surface vary from 12 to 5.6 kg/mm².

The constants of the material were as follows:

$$\sigma_b = 52.50 \quad \text{kg/mm}^2,$$
$$\sigma_{0.2} = 21.00 \quad \text{kg/mm}^2,$$
$$\sigma_{-1} = 24.50 \quad \text{kg/mm}^2.$$

No failure flaws were observed at 15,000 cycles, although the calculated stresses exceeded σ_b.

If we start with the elastic stress distribution alone, and take into consideration that the limiting value of endurance, corresponding with 15,000 cycles, is equal to 33.00 kg/mm², the safety factor is found to be:

$$n = \frac{1}{\dfrac{34.00}{52.50} + \dfrac{22.00}{33.00}} = 0.76.$$

Taking into consideration the stress reduction from plastic deformation, when $\sigma_m \approx 0$, we find the following value for the safety factor:

$$n = \frac{1}{\dfrac{0}{52.50} + \dfrac{22.00}{33.00}} = 1.5.$$

This serves to explain why failure did not occur at 15,000 cycles.

Langer proposes extending this method of calculation to the case of a compound stressed state including stress concentration. The following preliminary assumptions are recommended:

1. The stress intensities S must be used in the equation instead of the stresses, i.e., the strength for different stressed states is calculated on the Huber-Mises-Henky theory.

2. The effect of stress concentration is taken account of only in calculating the amplitude of the stresses.

The calculation is made in the following order:

1. The principal stresses σ_{1min} and σ_{2min} are found at the smallest temperature gradient.

2. The principal stresses σ_{1max} and σ_{2max} are found at the highest temperature gradient.

3. The corresponding characteristics σ_1 and σ_2 of the stress cycles are calculated

$$\sigma_{1m} = \frac{\sigma_1 \max + \sigma_1 \min}{2} ; \quad \sigma_{2m} = \frac{\sigma_2 \max + \sigma_2 \min}{2} ; \tag{41}$$

$$\sigma_{1a} = \frac{\sigma_1 \max - \sigma_1 \min}{2} ; \quad \sigma_{2a} = \frac{\sigma_2 \max - \sigma_2 \min}{2} . \tag{42}$$

4. The effective stress concentration factor, k_f, is found for each principal stress.

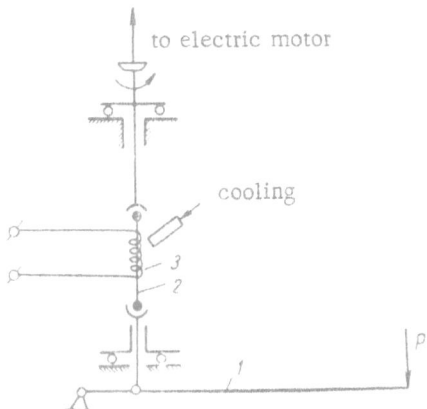

to electric motor

cooling

Fig. 61. Diagram of outfit for making thermal fatigue tests under static tension [71]: (1) lever; (2) sample; (3) inductance.

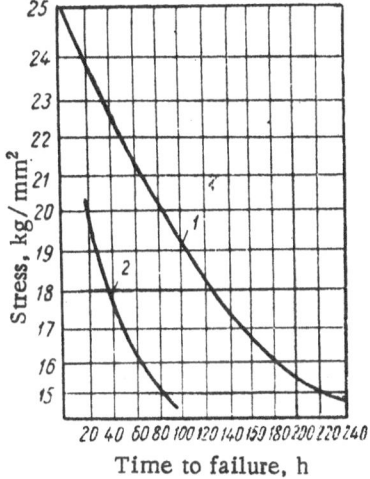

Fig. 62. Long-time strength of ÉI437 alloy [71]: (1) T = const = 800°C (without air cooling); (2) T_{max} = 800°C (with air cooling).

5. The mean stress intensity and the amplitude of the stress intensity are found:

$$S_m = \sqrt{\sigma_{1m}^2 + \sigma_{2m}^2 - \sigma_{1m} \sigma_{2m}}; \qquad (43)$$

$$S_a = \sqrt{(k_{1f} \sigma_{1a})^2 + (k_{2f} \sigma_{2a})^2 - k_{1f} \sigma_{1a} k_{2f} \sigma_{2a}}. \qquad (44)$$

If $S_m + S_a > \sigma_T$, $S_m + S_a$ must be reduced to the stress which is found on the deformation curve.

6. The safety factor is found from the expression

$$n = \frac{1}{\dfrac{S_a}{S_{-1}} + \dfrac{S_m}{S_b}}, \qquad (45)$$

where S_b is the ultimate strength, and S_{-1} is the limiting value of endurance in a symmetric cycle.

A number of investigators (see discussion to Langer's paper [70]), assume that it is a good idea in thermal fatigue to make the calculation from the deformations rather than from the stresses, since the fatigue process is primarily determined by the presence of alternating plastic deformations.

Note also that the diagram of the limiting cycles, $\sigma_m - \sigma_a$, used by Langer requires some refinement in view of the fact that the material is operating under alternating temperatures, while the constants σ_b and σ_{-1} are usually known for a given constant temperature.

Langer recommends taking account of the effect of stress concentrations by means of the coefficients k_f the values of which are unknown for large plastic deformations and so must be found from appropriate experiments.

Langer's method is a first, very approximate attempt to make a thermal fatigue calculation under simultaneous static loading.

Use of Tests

In finding the effect of repeated thermal stresses under simultaneous static loading, A. A. Klypin [71] used the outfit shown diagrammatically in Fig. 61.

The sample, 5 mm in diameter, preloaded by the force P to a stress of 0.6-4.5 kg/mm², was subjected to the effect of alternating temperatures, produced by induction heating, and was then cooled in a stream of water or air. In order to get uniform cooling on all sides, the sample was rotated by an electric motor. The grid circuit of the high-frequency generator as well as the cooling system were turned on and off automatically by two RVÉ-2 electronic time relays and two RP-0 interval relays. The materials studied were ÉI437 alloy and 1Kh18N9T steel.

The results of the tests on ÉI437 alloy are given in Fig. 62. The alloy samples were tested in two different ways: at a constant temperature of 800°C (without air cooling) and under alternating temperature with T_{max} = 800°C (with air cooling). The accelerated failure under repeated heating and cooling is accounted for by the author as being due to the rapid development of flaws resulting from the tensile thermal stresses produced on the surface when the sample is cooled.

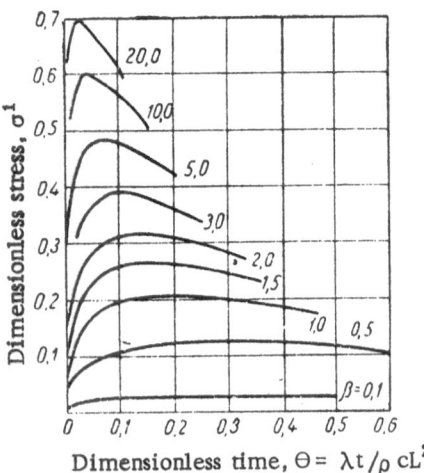

Fig. 63. σ^1 as a function of the time Θ
for different values of the heat transfer
parameter $\beta = Lh/\lambda$ [1].

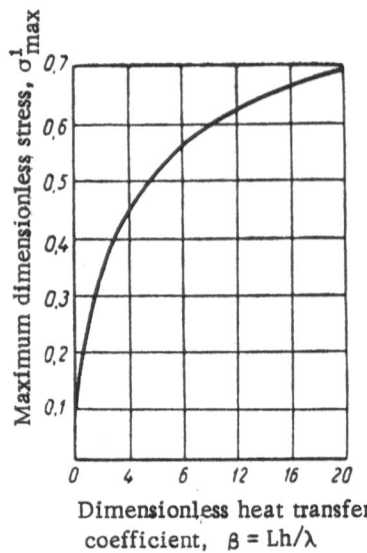

Fig. 64. σ^1_{max} as a function of β [1].

The results of these tests cannot be used to predict the behavior of the same material in parts of any other shape or size or under any other temperature cycles than those used in the tests, since the author did not determine the actual resulting stresses from static loading and thermal action. An analysis of these stresses, which could be represented by an asymmetric cycle, would permit a quantitative comparison of the strength to be made with different loading conditions under mechanical and thermal action.

STRENGTH UNDER THERMAL SHOCK

A sudden large change in temperature producing large thermal stresses is known as heat, or thermal, shock. The shock may be produced by a sudden temperature change in the ambient medium or by a change in internal heat production.

Thermal shock is characterized by large stresses due to the high temperature gradient, as well as by high rate of stress application which, under definite conditions, embrittles the material, and produces a dynamic effect as a result of accelerating the motion of various parts of the body.

Designing for Thermal Shock

A calculation of the strength of brittle materials under thermal shock has been made by S. Manson [1]. Since brittle failure (fracture) occurs in thermal shock, it is a good idea to take the maximum tensile stresses as the criterion of strength. Then the ultimate bending stress σ_b is found, where the stress distribution is very nearly the same as that observed in bodies undergoing thermal shock.

The laws governing the behavior of materials under thermal shock are derived from tests on a uniform flat plate which is initially at a uniform temperature and is then immersed in a medium at a lower temperature.

In order to find the stresses, we need to know the temperature distribution in the plate as a function of the time t after the temperature change occurs. For the case where the properties of the material are independent of temperature, and the material is perfectly elastic, we can write the following equation:

$$\sigma^1 = \frac{T_m - T}{T_0} , \tag{46}$$

where σ^1 is the ratio of the actual stresses σ to the stresses which would occur if deformations were completely prevented, T_m is the mean temperature over the cross section, T is the temperature at the point in question, and T_0 is the amount by which the initially uniform temperature of the plate exceeds the ambient temperature.

The formula for σ^1 is written as:

$$\sigma^1 = \frac{\sigma(1-\mu)}{E\alpha T_0}, \tag{47}$$

where μ is Poisson's coefficient, E is the modulus of elasticity, and α is the coefficient of linear expansion.

Figure 63 gives the results of calculations made using the dimensionless quantities σ^1, β, and θ, where σ^1 is the dimensionless stress on the surface, β is the dimensionless heat transfer parameter, $\beta = Lh/\lambda$, where L equals half the thickness of the plate, h is the heat transfer coefficient, and λ is the coefficient of thermal conductivity, θ is the dimensionless time, $\theta = \lambda t/\rho cL^2$, where t is the time, ρ is the density of the material, and c is the heat capacity.

The maximum dimensionless stress σ^1_{max} is shown in Fig. 64 as a function of the dimensionless heat transfer parameter β. The functional relationship may be expressed approximately by the following equation:

$$\frac{1}{\sigma^1_{max}} = 1.5 + \frac{3.25}{\beta}, \tag{48}$$

or more accurately

$$\frac{1}{\sigma^1_{max}} = 1.5 + \frac{3.25}{\beta} - 0.5 l^{-\frac{16}{\beta}}. \tag{49}$$

By using this approximate expression we can find the relation between σ^1_{max} and the physical properties of the material for the two cases of β small and β large.

For the most part, β is comparatively small, and we can then write

$$\frac{1}{\sigma^1_{max}} = \frac{3.25}{\beta}, \quad \text{or} \quad T_0 = \frac{\lambda\sigma}{E\alpha}\frac{3.25(1-\mu)}{Lh}. \tag{50}$$

At failure, $\sigma_{max} = \sigma_b$, and the limiting temperature change in thermal shock will be:

$$T_{0\,max} = \frac{\lambda\sigma_b}{E\alpha}\frac{3.25(1-\mu)}{Lh}. \tag{51}$$

Calculating $T_{0\,max}$ gives the temperature drop to produce failure, as well as the criterion to be used in evaluating materials for resistance to thermal shock.

Since Poisson's coefficient μ is the same for almost all materials, the value of $\lambda\sigma_b/Ea$ may be used as a criterion of the resistance to thermal shock. Table 11 gives the results of thermal shock tests made on a number of materials. The samples were heated in a furnace and then cooled in a stream of cold air. If, for a given furnace temperature, the samples held out for 25 cycles, the temperature was raised by 93.3°C, and the tests were repeated until failure occurred. The data of Table 11 indicate that the quantity $\lambda\sigma_b/Ea$ can actually be used as a criterion for comparing the behavior of different materials under thermal shock. Note that instead of the ultimate bending strength, use was made of the ultimate tensile strength, which could produce an appreciable error in making a quantitative evaluation of the materials.

For large values of β, $\sigma^1_{max} = 1$ and $\sigma_{max} = E\alpha T_0/(1-\mu)$.

This stress corresponds with preventing deformation in any direction parallel to the surface of the plate.

Thus, for large values of β, the maximum temperature drop is found from the expression

$$T_{0\,max} = \frac{\sigma_b}{E\alpha}(1-\mu). \tag{52}$$

In this case, the criterion for the behavior of the material under thermal shock is the quantity σ_b/E_α, which is independent of λ. A physical explanation of this fact may be found if we consider the possible ways in which β may be large (L and h large, or λ small). If L is large, the surface layers take on the temperature of the cooling

TABLE 12. Relation between the Properties of Materials and the Resistance to Failure in Thermal Shock [1]

Material	No. of thermal shock cycles at temperatures T, °C				Coefficient of linear expansion α, $\times 10^{-6}$ 1/°C	Coefficient of thermal conductivity λ, cal/h-cm·deg	Modulus of elasticity E, at 980°C, $\times 10^6$ kg/cm²	Ultimate tensile strength S at 980°C, kg/cm²	Thermal shock parameter, $\lambda S/\alpha E$
	980	1094	1210	1320					
A alloy*	–	–	–	–	15	172.2	0.07	2331	820,000
80% TiC + 20% Co	25	25	25	25**	10	295.0	4.2	2422	36,500
TiC	25	25	25	17	8.3	295.0	4.2	1204	21,900
BeO	25	3	–	–	9.3	128.0	2.99	434	4,270
ZrSiO$_4$	1	–	–	–	4.57	14.3	1.68	609	2,460
MgO	$\frac{1}{2}$	–	–	–	14	19.7-49.2	0.8	217	754-1,880
94% ZrO$_2$ + 6% CaO	0	–	–	–	10	17.6	1.75	472.5	1,000

*Not tested, but probably has better properties than any of the materials given.
**Failure absent.

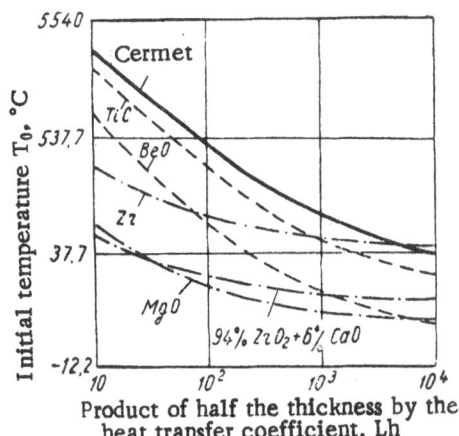

Fig. 65. Comparative behavior of materials under different conditions of thermal shock [1].

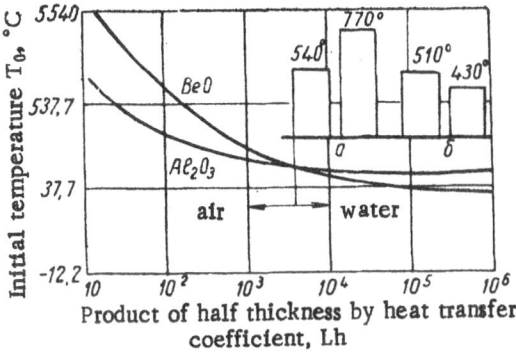

Fig. 66. Results of thermal shock tests on BeO and Al_2O_3 [1]: (a) failure temperature with air cooling; (b) failure temperature with water cooling.

medium up to the point where temperature changes occur in the mass of the solid, which practically eliminates the possibility of the surface layers being deformed. A similar picture is also observed for high values of h. If, however, λ is small, only the surface layers experience thermal shock.

The change in the criterion used to evaluate the behavior of metals under thermal shock, which occurs when the value of β changes, is of great importance in discussing the results of the tests. Figure 65 gives curves of $T_{0\,max}$ as a function of Lh (calculations made from the data of Table 12). It may be seen that for small values of Lh (~10), the values of $T_{0\,max}$ agree with the experimental data, according to which the strength of the materials come in the following order:

$$\text{cermet} > \text{TiC} > \text{BeO} > \text{Zr} > \text{MgO}$$
$$> 94\%\ \text{ZrO}_2 + 6\%\ \text{CaO}.$$

For large values of Lh, $T_{0\,max}$ differs from the experimental data. For example, for Lh > 8, zirconium is more stable to thermal shock than beryllium oxide. For very high values of Lh, the oxide behaves even worse than the other materials studied. The high thermal conductivity λ of beryllium oxide gives high resistance to thermal shock at low values of Lh, while for high values of Lh, the effect of high thermal conductivity is negligibly small.

Everything that has been said shows that the tests have to be made under conditions close to those actually existing. Trying to accelerate the tests by using more severe conditions in the experiment than exist in operation can lead to incorrect conclusions as to the thermal shock resistance of the material under actual conditions.

This is supported by the results of tests on Al_2O_3 and BeO made with two types of cooling – in air and in water. Figure 66 shows the maximum heating temperatures for these conditions, and curves of T_0 as a function of Lh, obtained by calculation.

Consider now a method of finding the thermal shock parameters experimentally.

The thermal shock resistance for the two experimental cases is expressed in terms of the following parameters:

$$P_1 = \lambda \sigma_b / E\alpha \text{ and } P_2 = \sigma_b / E\alpha. \tag{53}$$

From which

$$\lambda = P_1 / P_2 \text{ and } \beta = \frac{Lh}{\lambda} = \frac{LhP_2}{P_1}.$$

Then

$$\sigma^1_{max} = (1 - \mu) \frac{\sigma_b}{E\alpha T_0} = (1 - \mu) P_2 / T_{0\,max}. \tag{54}$$

111

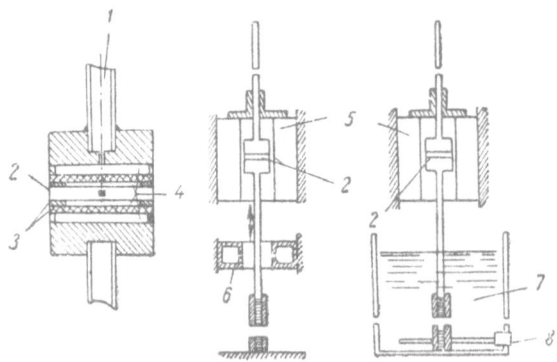

Fig. 67. Apparatus for measuring the thermal shock resistance
[1]: (1) thermocouple; (2) sample; (3) insulation; (4) air layer;
(5) furnace; (6) diffuser; (7) water; (8) heater.

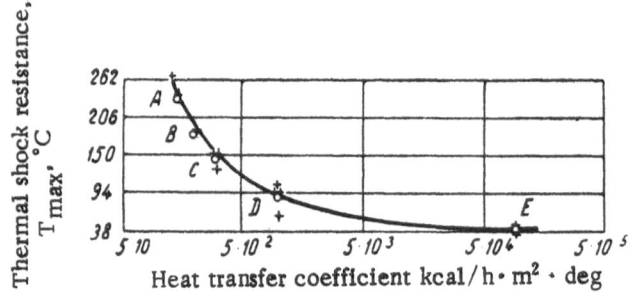

Fig. 68. Comparison between experimental data on thermal
shock resistance and calculated data based on measuring the
physical constants [$\lambda = 21.5$ cal/h·cm·deg, $\sigma_b/(Ea) = 39.4°C$]
(+) experimental data; (O) calculated data.

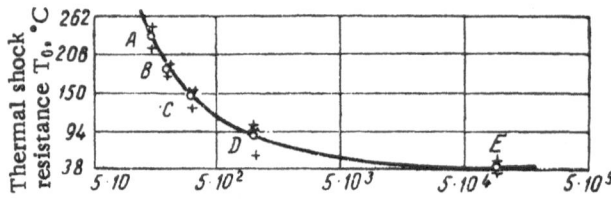

Fig. 69. Comparison of experimental thermal shock resistance
data with calculations based on a single test and a measurement
of the heat conductivity ($\lambda = 21.5$ cal/h·cm·deg, $\sigma_b/(E\alpha)$
$= 37.6°C$) [1].

Using Eq. (54), we can find the maximum temperature:

$$T_{0\,max} = (1 - \mu)\left[1.5 P_2 + \frac{3.25 P_1}{lh} - 0.5 P_2 l^{\frac{16 P_1}{LhP_2}}\right]. \tag{55}$$

112

Thus, T_0 may be calculated from the known values of P_1 and P_2.

That it is possible to make such calculations was verified from the results of tests on a ceramic material (ceramic steatite) for different cooling conditions, i.e., different values of h. The samples, in the form of discs, were subjected to thermal action on the cylindrical surface (the end surfaces were insulated) by heating in a furnace, followed by cooling in air or in water (Fig. 67). In the first case, the cooling was varied by changing the air pressure, and in the second, by means of a heater immersed in the water bath. The value of h was found from measuring the temperature of the disc.

The first method of finding the various parameters consisted of measuring the following physical quantities: λ, σ_b, E, and α. Using these quantities, Manson [1] drew a curve of $T_{0\,max}$ as a function of h (Fig. 68), which gives a comparison between the results of the calculations and the experimental data. The agreement between the calculated and experimental data must be regarded as very good. Here, in the expression for $T_{0\,max}$, the coefficient $(1 - \mu)$ was eliminated, since the tests were made on relatively thin discs.

The other method of calculation consists of making an experimental determination of the thermal shock resistance for a known value of β, and of measuring the thermal conductivity λ. Using these values, as well as the relation $P_2 = \lambda P_1$, the values of the parameters may be found. Figure 69 gives the curve of $T_{0\,max}$ as a function of h based on measurements, corresponding with point A, for which h = 150 kcal/h $\cdot$ m^2 $\cdot$ deg, $T_{0\,max}$ = 235°C, $\sigma_b/E\alpha$ = 37.7°C, and λ = 21.52 cal/h-cm-deg. In the same way as in the preceding cases, there is good agreement between the calculated and the experimental data. The best way of making the calculation is from measurements corresponding to large values of β, where $T_{0\,max}$ changes very little for large changes of β.

In making the calculations we have to keep in mind the relation between the physical properties of the material and the temperature, which changes within wide limits over the cross section of the sample.

All the above discussion is for thermal shock in which the heated body is cooled at the surface, where tensile stresses occur.

On the other hand, when a body is rapidly heated, compressive stresses occur in the surface layers, while tensile stresses occur inside.

If the compressive stresses are comparatively large, failure may occur on the surface by crumbling, as a result of transverse deformation, or by shear resulting from the maximum tangential stresses.

However, under certain conditions, failure can also occur inside the body as a result of tensile stresses. The maximum value of the tensile stress depends on the heating time and the thermal conductivity of the material. With poor conductivity, short time thermal shock is only transmitted to the surface layers, in which large compressive stresses occur for relatively small tensile stresses inside the body. If, however, the material is a good conductor of heat, the tensile stresses will be greater than the compressive stresses. Accordingly, under definite conditions, a material with good thermal conductivity will be less resistant to thermal shock than a material with poor thermal conductivity.

The magnitude of the stresses also depends on the duration of the heating, and it is thus not possible to find a simple parameter which gives the resistance of materials. Thus, in choosing the most suitable material, careful consideration must be given to the conditions under which it operates in the structure.

LITERATURE CITED

1. Manson, S., and Mach, S., Design 30(1958); No. 12-Appraisal of Brittle Materials; No. 13-Quantitative Techniques for Brittle Materials; No. 16-Basic Concepts of Fatigue in Ductile Materials; No. 17-Causes of Fatigue in Ductile Materials.
2. Serensen, S. V., and Kotov, P. I., Zavodsk. Lab. (9):1097 (1958).
3. Balandin, Yu. F., in: Metal Working, 3, Leningrad, Sudpromgiz, 1960.
4. Zalesskii, V. I., and Korneev, D. M., in: Production and Treatment of Steel, 32, Moscow Stal Institute, 1954, p. 237.

5. Fomin, S. F., Stal' (8):743 (1955).

6. Kostenko, D. I., Avtomob. i. Trakt. Prom. (8):29 (1957).

7. Holmberg, M., Trans. ASME 73(6):733 (1951).

8. Thielsch, H., Weld. Res. Council Bull. Ser. (10):April (1952).

9. Ratner, A. V., Izv. Vses. Teplotekhn. Inst. (10):12 (1948).

10. Khimushin, F. F., Heat Resistant Steels for Airplane Engines, Moscow, Oborongiz, 1942.

11. Prishchepa, M. P., Thermal Fatigue of Steels. Resume of Candidate's Dissertation, Tomsk, 1945.

12. Bochvar, A. A., and Novik, T. K., Dokl. Akad. Nauk SSSR 112(6):1041 (1957).

13. Sklyarov, N. M., et al., Zavodsk. Lab. (8):954 (1957).

14. Coffin, L., and Wesley, R., Trans. ASME 76(6):923 (1954).

15. Serensen, S. V., and Kotov, P. I., Zavodsk. Lab. (10):1216 (1959).

16. Kuznetsov, V. N., Teploénerg. (12):32 (1957).

17. Coffin, L., Proc. of Sagamore Conference, 1957.

18. Johansson, A., Colloquium on Fatigue, Berlin, 1956.

19. Baldwin, E., Sokol, G., and Coffin, L., Proc. ASTM 57:567 (1957).

20. Kennedy, C., Proc. Sagamore Conference, 1957.

21. Clauss, F., and Freeman, J., Thermal Fatigue of Ductile Metals, Part I, NACA Tech. Rept. 4160, Sept. 1958.

22. Kats, A. M., Theory of Elasticity, Moscow, Gostekhizdat, 1956.

23. Coffin, L., Trans. ASME 79(7):1637 (1957).

24. Glikman, L. A., Zhur. Tekhn. Fiz. 7(3):294 (1937).

25. Rädeker, W., Stahl u. Eisen 75(19):1252 (1955).

26. Liquid Metal Coolants (Sodium and Sodium – Potassium Alloy). Translation from English, edited by L. E. Sheidlin, Moscow, IL, 1958.

27. Holmberg, M., Welding J. 28(2):141 (1949).

28. Kostenko, D. I., Avtomob. i. Trakt. Prom.(26):30 (1957).

29. Muscatell, F., Reynolds, E., Dyrkacz, W., and Dolheim, J., Proc. ASTM 57:947 (1957).

30. Coffin, L., ASME, 1956, paper 56-A-178.

31. Coffin, L., Trans. ASME 78(3):527 (1956).

32. Griboedova, T. S., Metalloved. i Term. Obrabotka Metal (6):55 (1959).

33. Serensen, S. V., and Kozlov, L. A., Zavodsk. Lab. (11):1378 (1958).

34. Vitman, F. F., and Zlatin, N. A., Zhur. Tekhn. Fiz. 19:315 (1949).

35. Vitman, F. F., and Zlatin, N. A., Zhur. Tekhn. Fiz. 20:1267 (1950).

36. Nadai, A., and Manjoine, M., J. Appl. Mech. 8:A77 (1941).

37. Stepanov, V. M., Tr. LKVVIA (257):3 (1958).

38. Hunter, T., Symposium on Metallic Materials for Service at Temperatures Above 1600°F., June 1955, p. 164.

39. Haythorne, P., Iron Age, September, 89 (1948).

40. Pridantsev, M. V., and Krylova, A. R., Zavodsk. Lab. 24(2):204 (1958).

41. Maurer, E., and Haufe, W., Stahl u. Eisen 44 (51) (1924).

42. Lardge, H., Symposium on Metallic Materials for Service at Temperatures Above 1600°F, June 1955, p. 146.

43. Rädeker, W., Stahl u Eisen 74(15):929 (1954).

44. Masyutin, E. V., Increasing the Heat Resistance of Aircraft Alloys by Chemical and Heat Treatment. Resume of Candidate's Dissertation, LKVVIA, 1953.

45. Cranby, P., Metal Progr. 68(5):25 (1955).

46. Haufe, W., Stahl u. Eisen 47(33):1365 (1927).

47. Scherer, R., Stahl u Eisen 47 (48):2035 (1927).

48. Kishkin, S. T., and Klypin, A. A., Metalloved. i Term. Obrabotka Metal. (5):15 (1959).

49. Kurganov, G. V., and Sutina, Yu. A., Metalloved. i Term. Obrabotka Metal. (10):23 (1958).

50. Trusov, L. P. et al., Metalloved. i Term. Obrabotka Metal. (5):27 (1956).

51. Ratner, A. V., Teploénerg. (10):12 (1957).

52. Sobolev, N. D., Fiz. Metal. i Metalloved. 9(5):758 (1960).

53. Boas, W., and Honeycombe, R., Proc. Roy. Soc. A. 186(1004):57 (1946).

54. Boas, W., and Honeycombe, R., Proc. Roy. Soc. A. 188 (1015):28 (1947).

55. Wood, W., and Segall, R., J. Brit. Inst. Metals (Jan. 1958).

56. Fridman, Ya. B., and Egorov, V. I., Metalloved i Term. Obrabotka Metal. (7):27 (1960).

57. Boas, W., and Honeycombe, R., Nature 153(3886):494 (1944); 154(3906):338 (1944).

58. Likhachev, V. A., Inform. Byul. Leningr. Politekhn. Inst. (12):36, 44 (1958).

59. Chizuik and Kel'man, Effect of Cyclic Heat Treatment on Uranium, International Conference on the Peaceful Uses of Atomic Energy (Geneva, 1959). Vol. 9-Reactor Technology and Chemical Treatment of Nuclear Fuel, Moscow, Izd. AN SSSR, 1958, p. 184.

60. Avery, H., Trans. Am. Soc. Metals 38:957 (1947).

61. L'vovskii, M. Ya., and Smiyan, I. A., Zavodsk. Lab. 24(2):202 (1958).

62. Weisberg, H., and Soldan, H., Trans. ASME 76(7):1085 (1954).

63. Bentele, M., and Lowthian, J., Air Eng. 24(276):32 (1952).

64. Oding, I. A., and Kostochkin, Yu. V., Zavodsk. Lab. (7):863 (1959).

65. Fridman, Ya. B., Sobolev, N. D., and Egorov, V. I., Zavodsk. Lab. (4):467 (1960).

66. Coffin, L., Prod. Eng. 28 (6) (1957).

67. Haas, B., Arch. Eisenhüttenw. No. 5 (1956).

68. Kraston, D., et al., Atomnaya tekhnika za rubezhom, No. 4 (1954).

69. Shorr, B. F., Dokl. Akad. Nauk SSSR 123(5):809 (1958).

70. Langer, B., Trans. ASME 77, No. 5 (1955).

71. Klypin, A. A., The Phenomenon of Thermal Fatigue and the Failure Mechanism of Heat-Resistant Alloys at High Temperatures, Resume of Candidate's Dissertation MAI, Moscow 1954.

FUNDAMENTALS OF CREEP
CALCULATIONS ON NONUNIFORMLY HEATED PARTS

B. F. Shorr

1. Introduction

The development of engineering in recent years has been toward using higher and higher temperatures in power plants. Here, in addition to using heat-resistant materials, intense cooling of hot assemblies must be resorted to to ensure the strength of the structures. On the other hand, in the various types of heat exchange apparatus, the cooling of a number of parts is determined simply by the way in which the apparatus works. Since there is no heat transfer unless a temperature gradient is present, the temperature may differ by a large amount at different points in the parts being cooled, sometimes by hundreds of degrees.

Solids expand on heating, with the linear dimensions undergoing the relative elongation αt, where α is the coefficient of linear expansion, and t is the temperature change, and the volume of the solid undergoes a relative change equal to $3\alpha t$. In nonuniform heating, every particle in the solid tries to expand in accordance with its temperature. Neighboring particles which keep the solid together may prevent the expansion wholly or in part, with the result that temperature stresses occur in the solid, which is always internally in equilibrium, and hence statically indeterminate.

If the nonuniformity of the temperature field is relatively small, the thermal stresses only produce elastic deformations, so that after the body is cooled the stresses are completely removed. The methods of calculating thermal stresses in the elastic range are extensively treated in the literature [1-3].

In a large nonuniform temperature field, where the thermal stresses exceed the elastic limit, the material begins to be deformed plastically. On cooling in this case, residual stresses occur which can also reach a considerable magnitude. Only a very limited number of papers [4,5] have been devoted to calculating the temperature stresses in the elastoplastic deformation range.

Since, for the majority of structural materials, the coefficient of linear expansion is of the order of 10^{-5} per °C, the thermal deformations do not exceed fractions of a percent even for a temperature change of hundreds of degrees. Failure of plastic materials occurs at deformations measured in tenths of a percent, so that a single heating with even very large nonuniformity in the temperature field has practically no effect on the strength. Brittle materials (glass, ceramics) may fail in a single heating.

If the nonuniformly heated part undergoes vibration, the static thermal stresses, which change the asymmetry of the cycle, exert a definite effect on the fatigue failure of materials, including plastic materials. However, in the latter case, the effect is less for the reason that, because of plastic flow of the material, the temperature stresses in plastic solids do not exceed the flow limit, while in brittle solids the thermal stress concentration may be very high, especially in the surface layers.

If the heating and cooling is repeated, the alternating thermal stresses lead to the appearance and gradual development of fatigue cracks (like the ordinary effect of an alternating external load). The higher the level of the thermal stresses the smaller the number of cycles required to produce failure.

All these phenomena produced by thermal stresses (or better, by the expansion of solids when the temperature is raised) depend solely on the temperature difference in different parts of the solid, and have nothing to do with the actual values of the temperature. In speaking of the effect of nonuniform heating on strength, we often have in mind only the thermal stresses. But this is only true as long as the temperature in the hottest parts of the solid remains substantially below the melting point (approximately below $T_{max}/T_{mp} = 0.2-0.3$, where T is the absolute temperature, °K).

With further increase in temperature, as we analyze the effect of nonuniform heating on strength, other factors come to the fore which are related to the actual temperature in the solid and the way it changes throughout the volume. Such factors include: creep of the material, change in modulus of elasticity and deformation diagram with temperature, and change in the strength characteristics of the material with time and temperature.

The phenomenon of creep, as is well known, consists in the fact that when a solid is subjected to constant stresses, the plastic deformation increases with time. If the deformation does not encounter any obstacles, creep will cause a change in dimensions of the solid, but if the deformation of the whole or part of the solid is in some way limited, the increase in plastic deformation produces a reduction in elastic deformation, and hence a reduction in the stresses (stress relaxation occurs).

The creep rate increases rapidly with increase in stresses and particularly with increase in temperature. With a temperature difference of 100-200°, the creep rates in different parts of a solid experiencing stresses of the same order of magnitude may differ by factors of hundreds or even thousands.

In course of time, this leads to a considerable redistribution of the stresses from the external loads, the drop in stresses in the hot parts of the solid (where the creep rate is large) being made up for by the increase in the stresses in the colder parts. As far as the temperature stresses are concerned, since they are internally in equilibrium, they decrease throughout the whole volume of the solid, and tend gradually to disappear completely.

Thus, at high temperatures, the principal result of steady-state nonuniform heating is not so much the thermal stresses, as the stress distribution resulting from creep, accompanied by buildup in the plastic deformation.

Raising the temperature changes all the physical, mechanical, and strength characteristics of materials, which must be taken into account in making calculations, and decisive importance attaches to the length of time the part has been under load. At moderate temperatures, materials may be able to carry relatively small stresses for a practically unlimited time, but at high temperatures, stresses even considerably below the elastic limit lead to failure after a certain length of time (the phenomenon of "long-time" strength). The complicating factor in calculating the strength of nonuniformly heated parts is partly that the stresses change with time, while determinations of long-time ultimate strength are, as a rule, made under constant stresses.

The plastic deformations occurring in creep produce residual stresses in the solid after it is cooled, and these may be a cause of failure, particularly in materials that are inclined toward cold breakage. Under cyclic heating and cooling, the creep developed during each cycle gradually reshapes the stress cycles and changes the asymmetry of the cycle. This (in addition to other difficulties) makes it an unreliable business to use the results of ordinary fatigue strength tests in calculating the strength of materials under cyclic heating, and requires special tests for thermal stability to be made under conditions approximating those encountered in operation.

Nonuniform heating is also a direct or indirect cause of a number of other processes in structural elements operating at high temperature (formation of stresses of the second and third kind, the phenomenon of grain growth, various diffusional processes, etc.).

The problems of the strength of nonuniformly heated parts at high temperature are still in the initial stage of study at the present time. The present paper discusses only a few of the questions relative to creep calculations under nonuniform heating. First, a short analysis is given of creep theories, from the point of view of whether or not they apply to the present problem; then, in order of increasing complexity, the solution is given of a number of problems in creep calculation for nonuniformly heated rods, plates, and tubes, which are the basic elements used in building machinery.

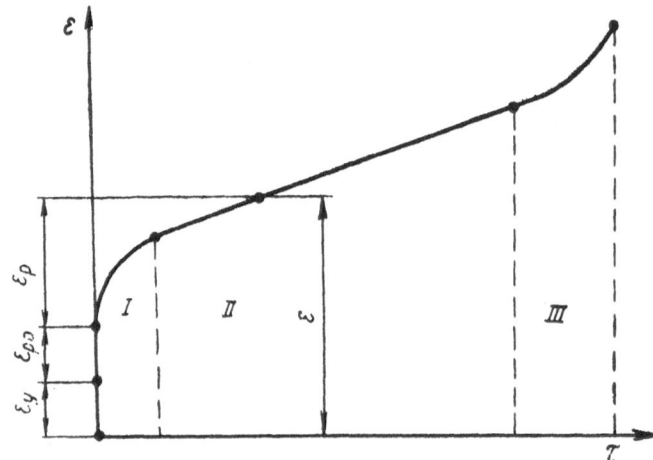

Fig. 1. Typical curve showing deformation of a sample
as a function of time at constant temperature.

2. Application of Creep Theories to Calculations on Nonuniformly Heated Parts

All the phenomenological theories of creep are based on experimental creep diagrams which give the deformations ε of the sample as a function of the time τ at constant temperature T and constant stress σ (Fig. 1). Tensile tests are the most widely used. The creep deformation is found as the difference between the total (geometric) deformation, $\varepsilon = \Delta l / l_0$, and the initial deformation produced when the sample is loaded, which in the general case may consist of two parts: an elastic part ε_e, and a plastic part ε_{p0}. It is important to note that the plastic creep deformation ε_p and the initial plastic deformation ε_{p0} are determined by different physical processes in a number of materials, so that their effect on hardening the material is also different [6].

As a rule, the creep rate $v_p = d\varepsilon_p / d\tau$ [1/sec] has its maximum value at the first instant of time, and then gradually decreases to some minimum value v_{min}, which may stay approximately constant for a considerable period of time. The range where the creep rate is decreasing (range I in Fig. 1) is called the unstabilized stage of creep, while the part with the constant rate v_{min} (range II) is called the stabilized stage. At moderate stresses, the unstabilized stage usually lasts 10 or 20 or more hours, so that it is the decisive factor in machines that operate for a short time, while in the design of equipment intended to be used for thousands and tens of thousands of hours, it is usually neglected. Creep is concluded in the third stage (range III) where local deformations (neckdowns) occur, accompanied by an increase in the rate v_p up to the instant of failure. As a rule, the third stage of creep should not be allowed to occur in structures.

The creep curves found at different stresses and temperatures give a definite function $F(\varepsilon_p, \sigma, T, \tau) = 0$ for a given material, which may be rewritten as the more general function $f(v_p, \varepsilon_p, \sigma, T, \tau) = 0$. A large number of methods have been proposed for expressing the functions F and f analytically (or graphically) [7].

We shall give a few of these:

Bailey [8]:

$$v_p = A\sigma^n; \quad A, \; n = f(T);$$

Kachanov [9]:

$$v_p = B(\tau)\sigma^n; \quad n = \text{const};$$

Malinin [10]:

$$\varepsilon_p = \psi\sigma, \quad \text{where} \quad \psi = \int_0^\tau B(\tau_1)\sigma^{n-1}(\tau_1)\,d\tau_1; \; n = \text{const};$$

118

Rabotnov [6]:

$$\varepsilon_p^\alpha v_p = A \exp (k\sigma - \beta/T); \quad A, \alpha, k, \beta = \text{const.}$$

All the methods of expressing the functions F and f, as well as the various methods of creep calculation based on them, may be classified in terms of a number of features, for example:

a. whether the basic function is the creep deformation ε_p or the rate v_p,

b. whether or not the theory is applicable to describing the unstabilized stage of creep, and if it is applicable, what is regarded as the fundamental variable determining the change in the creep rate – the time, or the accumulated plastic deformation (the first group includes the various aging theories, while the second includes theories of hardening).

In order to make an evaluation of creep theories from the point of view of whether or not they are applicable to the design of nonuniformly heated parts, we shall give the requirements which the theories must satisfy:

1. Since, in nonuniformly heated parts, the thermal stresses, and often the total stresses, are of opposite sign in different parts of the solid, with the signs of the stresses changing during creep even under constant external loads, the theory must hold under decreases in stresses, up to the point where they change sign.

2. Since thermal stress relaxation, which is most intense in the first hours of operation of the part, may, depending upon actual conditions, continue throughout the whole operating time, the theory must give a description of both unstabilized and stabilized creep.

3. In order to be able to analyze creep under repeated heating and cooling, the theory must be invariant as to the origin of the time scale.

4. The theory must be simple enough to use for practical engineering calculations.

In view of these requirements, obvious preference must be given to the theories which use the creep rate v_p as the basic parameter, rather than the deformation ε_p. The value of

$$\varepsilon_p = \int_0^\tau v_p (\tau_1) \, d\tau_1$$

will depend in an essential fashion on the entire "history" of the change with time in σ and T, and not merely on the values at any given instant.

Aging theories do not satisfy the third requirement above (they are not invariant to the origin of the time scale [11]), and are fundamentally unsatisfactory for giving a description of creep when the stresses change sign.

The requirements given are not fully satisfied by any of the known creep theories. The nearest thing to it is the theory of hardening, written in the form

$$\psi (v_p) = F (T, \sigma). \tag{2.1}$$

In the papers by Academician Yu. N. Rabotnov, the theory of strengthening (2.1), expressed in the form

$$v_p = A \exp \left(k\sigma - \frac{\beta}{T} \right) \varepsilon_p^{-\alpha}, \tag{2.2}$$

where A, k, β, and α are experimental constants, has been used to make an analysis of creep in nonuniformly heated solids [6]. However, the usefulness of Eq. (2.2) is limited to cases where the stresses in the solid are quite large ($\exp k\sigma \gg 1$), and do not change their sign with time. Further, Eq. (2.2) gives only the unstabilized stage of creep, and does not envisage a transition to the stabilized stage.

Hardening theory, which is suitable for giving a description of creep in a nonuniformly heated solid, and is free of the limitations mentioned, may be formulated for a uniaxially stressed state in the following way:

1. The direction of the creep rate is the same as the direction of the stress at any instant of time[*]:

$$\text{sign } v_p = \text{sign } \sigma. \tag{2.3}$$

2. The magnitude of the creep rate is determined by the values of the absolute temperature T and the stress σ at a given instant of time, and by the hardening function Φ, which gives the effect of loading history:

$$|v_p| = Ae^{-\beta(T)}[e^{k(T)|\sigma|} - 1]\Phi. \tag{2.4}$$

Here A = const, β, and k are experimental quantities, depending, in the general case, on the temperature [$\beta(T)$ decreases with increase in temperature]. Within certain limits of temperature change, we may assume k = const, and $\beta(T) \approx \beta/T$, where β = const. The sign $||$ indicates that the absolute value of the function is taken.

3. The fundamental characteristic of the loading history is taken to be the accumulated plastic deformation, calculated as[†]

$$p = \int_0^\tau |v_p(\tau_1)|\, d\tau_1. \tag{2.5}$$

In contrast with the instantaneous plastic deformation

$$\varepsilon_p = \int_0^\tau v_p(\tau_1)\, d\tau_1 \tag{2.6}$$

the accumulated deformation p is an essentially positive quantity, which increases as the creep develops.

If the sign of v_p does not change with time, $p = |\varepsilon_p|$.

4. Several types of creep may be considered, depending on the nature of the hardening function Φ:

a. Creep without hardening. The value of $|v_p|$ depends only on the instantaneous values of T and σ, i.e.,

$$\Phi = 1.$$

For T = const and σ = const, this corresponds with the stabilized stage of creep.

b. Creep with unlimited hardening, i.e., as the plastic deformation builds up, the magnitude of the creep $|v_p|$ decreases, approaching zero in the limit. For this case we set $\Phi = \Phi(p)$, for example, as in Yu. N. Rabotnov's paper [6],

$$\Phi = p^{-\alpha}, \tag{2.7}$$

where α is an experimental constant ($\alpha > 0$).

---

[*]The function sign f is defined in the following way:

$$\text{sign } f = \begin{cases} 1 & \text{for } f > 0, \\ 0 & , \quad f = 0, \\ -1 & , \quad f < 0. \end{cases}$$

[†]The value of p may be defined in more general form to include the directed nature of the hardening in creep [12], but no practical use can yet be made of this generalization, because of the lack of experimental data.

It is obvious that with the relation (2.7), Eq. (2.2) is a special case of Eq. (2.4), if sign v_p = const, and $\exp k\sigma \gg 1$.

According to the relation (2.7), at the first instant of creep (for p = 0), the magnitude of $\Phi(p)$, and hence of $|v_p|$, is infinitely large. This could be avoided by taking $\Phi(p) = (p + p_0)^{-\alpha}$, where p_0 is some new experimental constant. Calculations show, however, that finding the value of p_0 from creep diagrams is extremely inaccurate, and assuming p_0 = 0 does not appreciably change the results of the calculations. Accordingly, in what follows we shall take p_0 = 0.

c. Creep with limited hardening, i.e., hardening of type (2.7), occurs at small values of p, while for large values of p, the hardening stops. The boundary occurs at some value p_*, which depends on σ and T and is found from the creep diagram as the point of transition from the unstabilized to the stabilized stage. The hardening function may be written in the form

$$\Phi(p) = \left(\frac{p}{p_*}\right)^{-\alpha},\qquad (2.8)$$

where

$$\alpha = \begin{cases} \alpha(T,\ \sigma) & \text{for} \quad p/p_* \leqslant 1 \\ 0 & \text{for} \quad p/p_* > 1 \end{cases}\qquad (2.9)$$

The function $p_* = p_*(\sigma, T)$ is conveniently written in the form

$$p_* = A_* e^{-\beta_*(T)}\left[e^{k_*(T)|\sigma|} - 1\right].\qquad (2.10)$$

In this case, Eq. (2.4) satisfies the condition that at any instant of time the curve $\varepsilon_p = f(\sigma)$ is smooth, with no discontinuities [13], which is what is found experimentally.

For quite large stresses, where $\exp k|\sigma| \gg 1$, and $\exp k_*|\sigma| \gg 1$, it follows from Eqs. (2.4), (2.8), and (2.10) that

$$|v_p| = A_0 \exp\left[k_0(T)|\sigma| - \beta_0(T)\right] p^{-\alpha},\qquad (2.11)$$

where

$$\begin{aligned} A_0 &= AA_*^{\alpha}; \\ \beta_0 &= \beta + \alpha\beta_*; \\ k_0 &= k + \alpha k_*, \end{aligned}\qquad (2.12)$$

which, for $p < p_*$, and $\beta(T) = \beta/T$, gives Eq. (2.2).

Thus, the functions used in calculation contain seven independent experimental parameters found from the creep curves: the two constants A and A_*, and the five quantities β, β_*, k, k_*, and α, which depend in the general case on the temperature, but within definite limits may be replaced by their mean values.

3. Stabilized Creep of Nonuniformly Heated Parts

As was pointed out above, when samples are tested in tension, the creep rate becomes practically constant, starting at some instant of time, i.e., the creep passes to the stabilized stage. The stabilized creep rate is related to the stress and temperature by Eq. (2.4), with $\Phi = 1$.

For the creep to pass to the stabilized stage in the general case of the stressed state with arbitrary loading (i.e., for the creep rate to take some value independent of time at any point in the solid), the external load and the temperature must be constant (or become constant, starting at some instant of time). It is also obvious that the creep of the sample as a whole cannot become stabilized until a definite deformation $p \geq p_*$ has been built up at all points, and the hardening function has taken on the value $\Phi = 1$. However, these conditions are sufficient only if thermal stresses are absent.

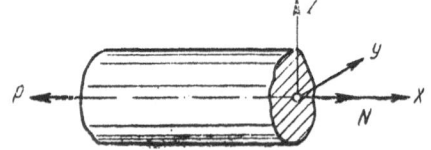

Fig. 2. Extensible rod.

The thermal stresses, being internally in equilibrium, relax in course of time at a gradually decreasing rate and theoretically only disappear completely as $\tau \to \infty$. However, for practical purposes, some finite value of the time may always be found, starting at which the residual thermal stresses no longer need to be taken into consideration. After this instant, the creep is able to pass to a practically stabilized stage.

Structural parts, designed for a long service life under constant loads and temperatures, may operate under conditions of stabilized creep for the greater part of the time. The long-time strength of such parts cannot be found from the short-time initial stresses, but from the stresses which exist for a long period of time, and are calculated in accordance with stabilized creep theory.

Deformation of a Rod Without Bending

Let a rod of arbitrary cross section with the area F be in tension under longitudinal forces which produce the internal forces N(x) in the cross sections (Fig. 2). The steady-state temperature distribution over the cross section is given in the form T = T(y,z). A temperature field of this sort may be produced by having longitudinal cooling channels in the rod or by having heat sources distributed over the cross section. The temperature does not change along the length of the rod.

Assume first that the cross section has two axes of symmetry, and that the temperature field is also symmetric with respect to the axes.

For the case where the rod is quite long in comparison with its cross-sectional dimensions, the cross sections at some distance from the ends will remain plane and perpendicular to the axis of the rod under deformation, i.e., the relative geometric elongation of all the fibers in the rod is the same:

$$\varepsilon = \text{const.} \tag{3.1}$$

At any instant of time, and at any point, the value of ε may be represented as the sum:

$$\varepsilon = \frac{\sigma}{E} + \varepsilon_{p0} + \varepsilon_p + \alpha t, \tag{3.2}$$

where σ/E is the elastic deformation, E is the modulus of elasticity, ε_{p0} is the rapid plastic deformation, as given by the "stress — deformation" diagram, ε_p is the creep deformation, α is the coefficient of linear expansion, and t is the temperature difference (for example, t = T − 293, where T is the absolute temperature, and the initial temperature is taken as T_0 = 20°C = 293°K).

Differentiating Eq. (3.2) with respect to the time, and bearing in mind that the quantities σ, ε_{p0}, and t are independent of time in stabilized creep, we obtain

$$\frac{d\varepsilon}{d\tau} = \frac{d\varepsilon_p}{d\tau}, \tag{3.3}$$

from which, using condition (3.1), it follows that the stabilized creep velocity of all the points in the rod under tension will be the same

$$v_p = \frac{d\varepsilon_p}{d\tau} = \text{const.} \tag{3.4}$$

Solving Eq. (2.4) for $|\sigma|$ with $\Phi = 1$, we find

$$|\sigma| = \frac{1}{k(T)} \ln \left[\frac{1}{A} |v_p| e^{3(z)} + 1 \right]. \tag{3.5}$$

If v_p = const, the sign of v_p, and thus the sign of σ do not change over the cross section. Since, from the equilibrium condition for any element of the rod

$$\int_F \sigma dF = N, \tag{3.6}$$

where the integral is taken over the whole area of the cross section, the sign of σ will be the same as the sign of the force N, which makes it possible to write, in place of Eq. (3.5),

$$\sigma = \frac{\text{sign } N}{k(T)} \ln \left[ae^{\beta(T)} + 1 \right]. \tag{3.7}$$

The constant positive quantity $a = |v_p|/A$ must be taken such as to satisfy condition (3.6). In the general case it is found by graphical analysis, for which purpose a graph is plotted of the function

$$\varphi(a) = N - \text{sign } N \int_F \frac{1}{k(T)} \ln \left[ae^{\beta(T)} + 1 \right] dF,$$

and the value of a is found at which $\varphi(a) = 0$.

Consider two special cases:

1. At all points in the cross section, $ae^{\beta(T)} \gg 1$ (or $e^{k|\sigma|} \gg 1$). Then it follows from Eqs. (3.6) and (3.7) that

$$\int_F \sigma dF = \text{sign } N \left[\ln a \int_F \frac{dF}{k(T)} + \int_F \beta(T) dF \right] = N,$$

from which

$$\ln a = k_m |\sigma_m| - \beta_m,$$

where

$$\sigma_m = \frac{N}{F}, \quad \frac{1}{k_m} = \frac{1}{F} \int_F \frac{dF}{k(T)}, \quad \beta_m = \frac{1}{F} \int_F \beta(T) dF.$$

Finally,

$$\sigma = \frac{k_m}{k(T)} \sigma_m + \frac{\text{sign } \sigma_m}{k(T)} [\beta(T) - \beta_m]. \tag{3.8}$$

2. At all points in the cross section, $ae^{\beta(T)} \ll 1$ (or $k|\sigma| \ll 1$). Using the expansion $\ln(1+x) = x$, which holds for $x \ll 1$, we find from Eq. (3.7)

$$\sigma = \text{sign } Naf(T),$$

where

$$f(T) = \frac{1}{k(T)} e^{\beta(T)}.$$

Further,

$$\int_F \sigma dF = \text{sign } Na \int_F f(T) dF = N,$$

from which

$$a = \frac{|N|}{\int_F f(T)\,dF}.$$

We obtain finally

$$\sigma = \sigma_m \frac{f(T)}{f_m}, \tag{3.9}$$

where

$$f_m = \frac{1}{F} \int_F f(T)\,dF.$$

Equations (3.7)-(3.9) make it possible to find the following general features of the stress distribution in a rod in tension (or compression) under stabilized creep conditions:

a. in a uniform temperature field, the stresses are constant over the cross section, and equal to the mean value σ_m;

b. in a nonuniform temperature field, the stresses drop off at the hotter points, and increase at the colder points (i.e., the nonuniform stressed distribution is equalized by the difference in creep rates produced by the nonuniform heating);

c. the stresses have the same sign at all points in the cross section; hence, it follows, in particular, that if there is no external force, the stresses are equal to zero (this once again confirms the fact that the thermal stresses are completely relaxed in stabilized creep);

d. for a small amount of stress, the relative nonuniformity σ/σ_m depends only on the temperature field and is independent of the absolute value of σ_m, while for a large amount of stress the relative nonuniformity decreases with increase in σ_m and, as $\sigma_m \to \infty$, $\sigma/\sigma_m \to 1$.

The way the absolute nonuniformity $\sigma - \sigma_m$ varies with the stress is opposite to this: the difference $\sigma - \sigma_m$ is proportional to σ_m for small stresses, while for large stresses it depends only on the temperature field.

It was assured above that the cross section of the rod has two axes of symmetry, and that the temperature field is also symmetric about these axes. If we abandon these assumptions, we have to deal with the possibility that the cross sections, while remaining plane, may rotate with respect to one another, i.e., a tensional deformation will occur accompanied by bending (see next paragraph). However, if the longitudinal force is applied at some particular point i which may for the present be called a longitudinal stiffness pole in creep, the condition (3.1) remains in force, and the rod will experience purely tensional deformation.

The coordinates of the longitudinal stiffness pole y_i, z_i may be found from the condition that the resultant of all the internal forces, σdF, must pass through this point, the function $\sigma = \sigma(y,z)$ being given by Eqs. (3.7), (3.8), or (3.9). In the general case

$$\left.\begin{aligned}
y_i &= \frac{1}{|N|} \int_F \frac{y}{k(T)} \ln\left[ae^{\beta(T)} + 1\right] dF, \\
z_i &= \frac{1}{|N|} \int_F \frac{z}{k(T)} \ln\left[ae^{\beta(T)} + 1\right] dF.
\end{aligned}\right\} \tag{3.10}$$

For $e^{k|\sigma|} \gg 1$, and $k = \text{const}$

$$\left.\begin{aligned}
y_i &= \frac{1}{k|N|} \int_F y\beta(T)\,dF, \\
z_i &= \frac{1}{k|N|} \int_F z\beta(T)\,dF.
\end{aligned}\right\} \tag{3.11}$$

124

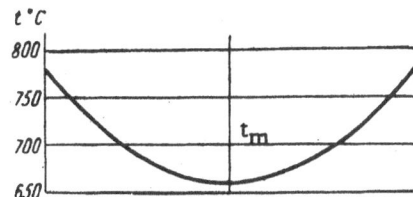

Fig. 3. Temperature distribution over the thickness of the rod.

For $k|\sigma| \ll 1$

$$
\left.
\begin{aligned}
y_i &= \frac{1}{Ff_{\text{m}}} \int_F y f\,(T)\,dF, \\
z_i &= \frac{1}{Ff_{\text{m}}} \int_F z f\,(T)\,dF.
\end{aligned}
\right\} \tag{3.11'}
$$

It is assumed in Eqs. (3.11) that the X and Y axes pass through the center of gravity of the cross section, i.e.,

$$
\int_F y\,dF = \int_F z\,dF = 0.
$$

It follows from Eqs. (3.10) and (3.11), as well as from simple physical considerations, that the pole i is located in the colder part of the cross section. By having the tensile force N directed along an axis passing through this point, the rod can be made to stay straight in stabilized creep for a long time. Having the longitudinal force located this way in compression will give the best resistance to lateral buckling.

The effect of curvature of the rod axis under elastic and plastic deformation in the unstabilized creep stage is not taken into consideration here.

The elongation of the rod in the stabilized stage increases linearly with the time τ, and may be found from the formula

$$
\varepsilon = \text{sign}\ \sigma_{\text{m}}\, A a \tau, \tag{3.12}
$$

while, if Eq. (3.8) applies,

$$
\varepsilon = \text{sign}\ \sigma_{\text{m}}\, A \exp\left[k|\sigma_{\text{m}}| - \frac{1}{F} \int_F \beta\,(T)\,dF \right] \tau.
$$

Example 1. Find the stress distribution in a rod with a rectangular cross section of area 1 cm², in tension under a force of 2.5 tons ($\sigma_{\text{m}} = 25$ kg/mm²). The material of the rod is a heat-resistant alloy which, in the temperature range 600-800°C, has the averaged stabilized creep constants

$$
k = 0.2854 \text{ mm}^2/\text{kg}
$$

$$
\beta = 5.34 \cdot 10^4\ ^\circ\text{K} \left(\text{for}\quad \beta\,(T) = \frac{\beta}{T} \right).
$$

The temperature distribution over the thickness of the rod is parabolic, with the mean temperature $t_{\text{m}} = 700°C$, and the nonuniformity (maximum temperature minus minimum temperature) $\Delta t = 120°C$ (Fig. 3). The temperature is constant over the width of the rod. It is assumed that the rod is in the stabilized creep state.

Solution. The distribution of absolute temperatures is given by the formula

$$
T\,(\bar{y}) = (t_{\text{m}} + 273) - \frac{\Delta t}{3}\,(1 - 12\,\bar{y}^2),
$$

where $\bar{y} = y/H$, and H is the thickness of the rod.

In making the calculations, it is a good idea to introduce the auxiliary constant $a_1 = ae^{\beta/T_{\text{max}}}$, and the function $\beta_1/T = \beta/T - \beta/T_{\text{max}}$, giving

$$
ae^{\beta\,(T)} = ae^{\beta_1/T + \beta/T_{\text{max}}} = a_1 e^{\beta_1/T},
$$

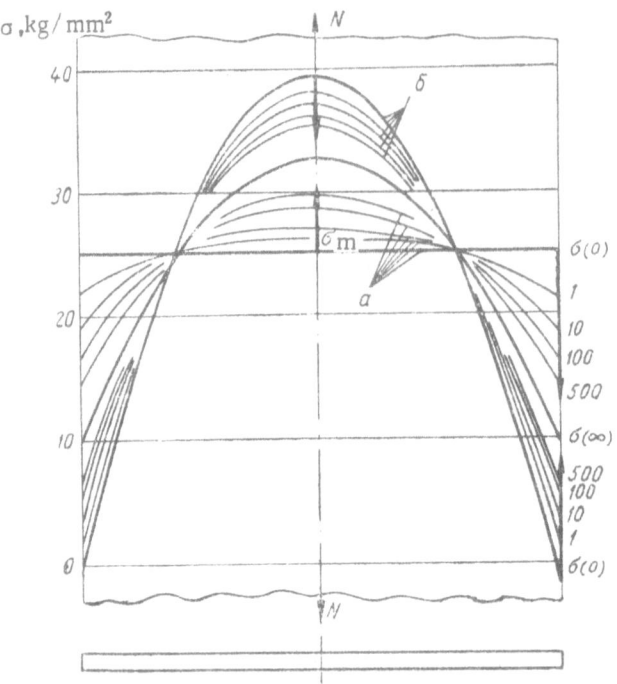

Fig. 4. Stress distribution over the thickness of a nonuniformly heated rod in tension at different instants of time: (a) neglecting thermal stresses; (b) including thermal stresses. The figures at the right give the time in hours.

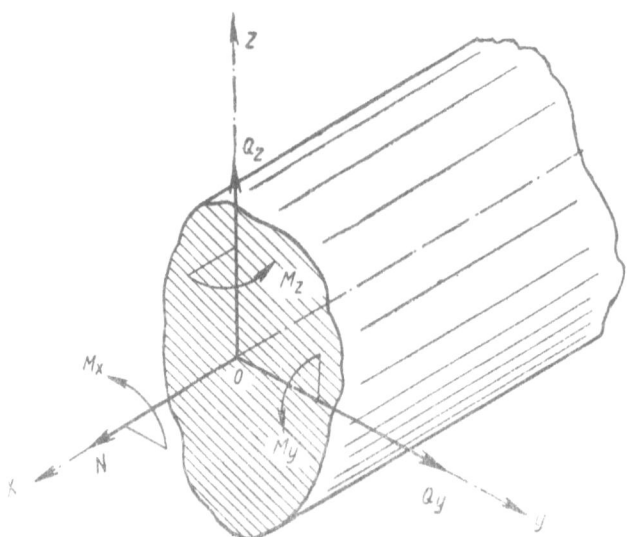

Fig. 5. Force factors in the cross section of a rod in the general loading case.

and

$$\ln (ae^{\beta/T}) = \ln a_1 + \beta_1/T.$$

By calculating the function

$$\varphi(a_1) = \sigma_m - \frac{2}{k} \int_0^{1/2} \ln [a_1 e^{\beta_1/T\,(\overline{y})} + 1]\, d\overline{y},$$

for several values of a_1, we find that $\varphi(a_1) = 0$ for $a_1 = 17.7$.

With the approximate formula

$$\ln a_1 = k\sigma_m - 2 \int_0^{1/2} \frac{\beta_1}{T\,(\overline{y})}\, d\overline{y},$$

we also find that $a_1 \approx 17.7$.

The stress distribution calculated from Eq. (3.7) is shown in Fig. 4, where it is designated as $\sigma(\infty)$. A calculation from Eq. (3.8) gives the same curve. Equation (3.9) obviously does not apply in this case.

Deformation of a Rod with Bending

Consider a rod of arbitrary cross section, loaded by some system of forces. In the general case we have acting in the cross section of the rod: the longitudinal force N, the transverse forces Q_y and Q_z, the bending moments M_y and M_z, and the torsional moment M_x (Fig. 5).

We assume that there is no torsional moment M_x, and that the effect of the transverse forces Q_y and Q_z on the deformation of the rod may be neglected. Then, for parts of the rod sufficiently far from the ends, we may assume that the plane-cross-section hypothesis holds, according to which the relative geometric elongation in the longitudinal direction is given by the linear equation

$$\varepsilon = \varepsilon_0 - \varkappa_y y - \varkappa_z z, \tag{3.13}$$

where ε_0 is the elongation of a fiber passing through the center of gravity of the cross section, $\varkappa_y$ and $\varkappa_z$ are the curvatures of the fiber in the xy and xz planes, respectively, and y and z are the coordinates of a point in the cross section in any perpendicular axes passing through the center of gravity of the cross section.

The temperature distribution over the cross section, T (y,z), is given.

Since, for stabilized creep in the uniaxial stressed state, Eq. (3.2) will hold as before, setting Eq. (3.2) equal to (3.13) and differentiating the resulting equation with respect to the time, we obtain (for constant values of σ, ε_{p0}, and T)

$$v_p = \frac{d\varepsilon_0}{d\tau} - \frac{d\varkappa_y}{d\tau} y - \frac{d\varkappa_z}{d\tau} z, \tag{3.14}$$

i.e., in stabilized creep, the rate is distributed over the cross section in accordance with the plane-cross-section hypothesis.

It follows from the basic relations (2.3) and (2.4), with $\Phi = 1$, that

$$\sigma = \frac{\operatorname{sign} v_p}{k\,(T)} \ln \left[\frac{|\,v_p\,|}{A} e^{\beta\,(T)} + 1 \right]. \tag{3.15}$$

Putting the value of v_p from Eq. (3.14) into Eq. (3.15), and using the notation

$$a = \frac{1}{A} \frac{d\varepsilon_0}{d\tau}, \quad b = -\frac{1}{A} \frac{d\varkappa_y}{d\tau}, \quad c = -\frac{1}{A} \frac{d\varkappa_z}{d\tau},$$

we obtain

$$\sigma = \frac{1}{k(T)} \, \text{sign} \, (a + by + cz) \ln \left[\, |\, a + by + cz \,|\, e^{\beta(T)} + 1 \right].$$ (3.16)

The constants a, b, and c are found from the three equilibrium equations

$$
\left.
\begin{aligned}
\int_F \sigma \, dF &= N, \\[4pt]
\int_F \sigma_y \, dF &= -M_z, \\[4pt]
\int_F \sigma_z \, dF &= M_y.
\end{aligned}
\right\}
$$ (3.17)

It is a good idea to make the calculation in the following order. Taking at the start $b^{(0)} = c^{(0)} = 0$, we set up the function

$$\varphi(a) = N - \int_F \sigma(a) \, dF$$

and find $a^{(1)}$ from the condition $\varphi(a^{(1)}) = 0$. Further, we set up the function

$$\psi(b) = -M_z - \int_F \sigma(b) \, y \, dF,$$

where, in calculating $\sigma(b)$, we assume $a = a^{(1)}$, and $c^{(0)} = 0$. Finding $b^{(1)}$ from the condition $\psi(b^{(1)}) = 0$, we calculate $c^{(1)}$ in a similar way, and then in the same order correct the values of the constants in the second approximation. If one of the force factors obviously predominates over the others, it is a good idea to begin the calculation with the constant corresponding to this force factor.

It may be seen from Eq. (3.16) that in stabilized creep of a nonuniformly heated rod the stresses are distributed over the cross section in a complicated way. However, the points at which the stresses are equal to zero are located, as in bending under ordinary conditions, on a straight line (neutral line), the equation of which is

$$a + by + cz = 0.$$ (3.18)

Obviously, the creep rate of the points on a neutral line is also equal to zero.

In the special case of plane bending (in the xy plane), we have c = 0, and then

$$\sigma = \frac{1}{k(T)} \, \text{sign} \, (a + by) \ln \left[\, |\, a + by \,|\, e^{\beta(T)} + 1 \right].$$ (3.19)

If the temperature is constant over the cross section, the quantity $e^{\beta(T)}$ may be included in the corresponding constants which, instead of Eq. (3.19), gives the relation

$$\sigma = \frac{1}{k} \, \text{sign} \, (a' + b'y) \ln \left[\, |\, a' + b'y \,|\, + 1 \right],$$ (3.20)

i.e., the stresses in this case are distributed logarithmically over the cross section.

Consider in more detail the practically interesting case of plane bending of a rod the cross section and temperature field of which are symmetric about the Y and Z axes.

Here, a = c = 0, and

$$\sigma = \frac{1}{k(T)} \, \text{sign} \, (by) \ln \left[\, |\, by \,|\, e^{\beta(T)} + 1 \right].$$ (3.21)

Substituting Eq. (3.21) in (3.17) gives sign b = −sign M_z, so that

$$\sigma = -\frac{1}{k(T)} \operatorname{sign}(M_z y) \ln\left[|by|e^{\beta(T)} + 1\right].$$

(3.22)

As in the tension problem, we shall consider two special cases:

1. Large bending stresses occur over the whole cross section, with the exception of the very narrow zone adjacent to the neutral line, where the stresses are small.

In this case, Eq. (3.22) may be replaced approximately by the following:

$$\sigma = \begin{cases} -\dfrac{1}{k(T)} \operatorname{sign}(M_z y)\left[\ln|b| + \ln|y| + \beta(T)\right] & \text{for} \quad |by|e^{\beta(T)} \geqslant 1, \\ 0 & \text{for} \quad |by|e^{\beta(T)} < 1. \end{cases}$$

(3.23)

Further,

$$-M_z = -2\operatorname{sign} M_z \left\{ \ln|b| \int_{F/2} \frac{y\,dF}{k(T)} + \int_{F/2} \frac{y\ln y}{k(T)}\,dF + \int_{F/2} \frac{y\beta(T)}{k(T)}\,dF \right\},$$

from which

$$\ln|b| = \frac{k_m|M_z|}{S_z} - \frac{S_z^{\ln}}{S_z} - \frac{S_z^{\beta}}{S_z},$$

(3.24)

where

$$S_z = k_m \int_{F/2} \frac{y\,dF}{k(T)},$$

$$S_z^{\ln} = k_m \int_{F/2} \frac{y\ln y\,dF}{k(T)},$$

$$S_z^{\beta} = k_m \int_{F/2} \frac{y\beta(T)\,dF}{k(T)}.$$

For k = const, the value of S_z is the same as the static moment of half the cross section with respect to the Z axis.

Strictly speaking, the integrals in Eq. (3.24) need to be evaluated only for the part of the cross section where $(by)e^{\beta(T)} \geq 1$. However, if the zone along the neutral line, where the above inequality is not satisfied, is small, the effect of the zone may be neglected altogether.

Substituting the value of $\ln|b|$ in Eq. (3.23), we obtain

$$\sigma = \begin{cases} -\operatorname{sign} y \left\{ \dfrac{k_m M_z}{k(T)\,2S_z} - \dfrac{\operatorname{sign} M_z}{k(T)} \left[\left(\dfrac{S_z^{\ln}}{S_z} - \ln|y| \right) \right. \right. \\ \left. \left. \qquad + \left(\dfrac{S_z^{\beta}}{S_z} - \beta(T) \right) \right] \right\} \text{ for } \quad |by|e^{\beta(T)} \geqslant 1, \\ 0 \qquad\qquad\qquad\qquad\quad \text{for} \quad |by|e^{\beta(T)} < 1. \end{cases}$$

(3.25)

If the creep rate increases very strongly with increase in stress, the coefficient k has a large value, and if we take $k \to \infty$ in the limit, we obtain from Eq. (3.25)

$$\sigma = -\operatorname{sign} y \frac{M_z}{2S_z},$$

(3.26)

which corresponds with the stress distribution for ideal plasticity. For any actual values of k, the stresses change in accordance with Eq. (3.25). It is obvious that Eq. (3.26) is also a limiting form of (3.25), as the bending moment M_z becomes large.

2. For small values of the moment M_z, where $[\,|\,by\,|\,e^{\beta(T)}]_{max} \ll 1$, it follows from Eq. (3.22) that:

$$\sigma = - \frac{\text{sign } M_z \quad b\,|\,ye^{\beta(T)}}{k(T)}. \tag{3.27}$$

Putting (3.27) into the equilibrium equation (3.17), we find

$$|\,b\,| = \frac{|\,M_z\,|}{I_z^\beta},$$

where

$$I_z^\beta = \int\limits_F y^2 f(T)\,dF, \quad f(T) = \frac{e^{\beta(T)}}{k(T)},$$

so that

$$\sigma = - \frac{M_z y f(T)}{I_z^\beta}. \tag{3.28}$$

In a uniform temperature field, the distribution of the bending stresses changed as a function of the quantity $k\,|\,\sigma_{max}\,|$. For $k\,|\,\sigma_{max}\,| \ll 1$, the distribution is of a linear nature, and then as $k\,|\,\sigma_{max}\,|$ increases, it changes to a logarithmic curve in accordance with (3.25) and, in the limit, as $k\,|\,\sigma_{max}\,| \to \infty$, the stresses in the positive and negative regions become constant.

A nonuniform temperature distribution distorts this regularity with, as in tension, the stresses dropping off in the hotter parts of the cross section, and rising in the colder parts.

To find the deformations in bending, we must go from the constants a, b, and c to the corresponding derivatives

$$\frac{d\varepsilon_0}{d\tau} = Aa, \quad \frac{d\varkappa_y}{d\tau} = -Ab, \quad \frac{d\varkappa_z}{d\tau} = -Ac,$$

which, after integrating over the time, and employing the usual geometric relations

$$\varepsilon_0 = \frac{du}{dx}, \quad \varkappa_y = \frac{d^2v}{dx^2}, \quad \varkappa_z = \frac{d^2w}{dx^2},$$

where u, v, and w are the displacements along the X, Y, and Z axes, gives

$$\left.\begin{aligned}
\frac{du}{dx} &= Aa\tau, \\
\frac{d^2v}{dx^2} &= -Ab\tau, \\
\frac{d^2w}{dx^2} &= -Ac\tau.
\end{aligned}\right\} \tag{3.29}$$

The integration of Eqs. (3.29) over the length of the beam is carried out by the usual methods.

Example 2. Find the stress distribution in a rod of rectangular cross section with the dimensions B = 1 cm, H = 6 cm, acted upon by the bending moment M_z = 15 ton-cm. The material of the rod is the same and has the same characteristics as in Example 1. The temperature distribution over the thickness is linear, with the mean temperature t_m = 700°C, and the nonuniformity Δt = 120°C (Fig. 6). The rod is in the stabilized stage of creep. Compare with the data of an "elastic" calculation and with a creep calculation for the mean temperature.

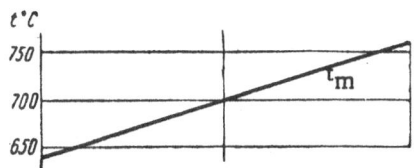

Fig. 6. Temperature distribution
over the thickness of the rod.

From an elastic calculation,

$$\sigma_{max} = \frac{6M_z}{BH^2} = \frac{6 \cdot 15\,000}{1 \cdot 6^2} = 2500 \text{ kg/cm}^2.$$

The thermal stresses in this case are equal to zero (if the edges of the rod are free).

In the second case, the distribution of absolute temperatures is given by the formula

$$T(\bar{y}) = (t_m + 273) + \Delta t \bar{y}, \quad (\bar{y} = y/H),$$

and, in the third case, T = const.

The stress distribution calculated for all cases from Eq. (3.19) is shown in Fig. 7, from which it may be seen that the curve of the stresses σ_∞ for stabilized creep are greatly different from the data σ_0 of the elastic calculation, as well as from the creep calculation made at the mean temperature, $\sigma_\infty (t_m)$.

Stabilized Creep in a Compound Stressed State

In the general case, the stressed state at a point is determined by the three principal stresses σ_1, σ_2, and σ_3, and may be represented as the sum of two stressed states — uniform tension (compression) on all sides, with the stress

$$\sigma = \frac{1}{3}(\sigma_1 + \sigma_2 + \sigma_3) \tag{3.30}$$

and the generalized shear, the principal stresses of which are equal to:

$$\sigma_1' = \sigma_1 - \sigma, \quad \sigma_2' = \sigma_2 - \sigma, \quad \sigma_3' = \sigma_3 - \sigma.$$

In exactly the same way, the creep deformation rate of a given element in the solid is determined by the three principal deformation rates v_{p1}, v_{p2}, and v_{p3}, and may be represented as the sum of two deformation rates — the rate of change in volume, equal to

$$v_p = v_{p1} + v_{p2} + v_{p3}, \tag{3.31}$$

and the rate of shear deformation with the principal components

$$v_{p1}' = v_{p1} - \frac{1}{3}v_p, \quad v_{p2}' = v_{p2} - \frac{1}{3}v_p, \quad v_{p3}' = v_{p3} - \frac{1}{3}v_p.$$

Experiments show that plastic deformation occurs with practically no change in volume, so that $v_p = 0$, and

$$v_{p1}' = v_{p1}, \quad v_{p2}' = v_{p2}, \quad v_{p3}' = v_{p3}.$$

If we consider not the principal, but rather, arbitrary directions, the stresses and deformation rates will change with the direction. However, the values of σ and v_p will remain unchanged, since they are independent of the choice of coordinate systems. It may be shown that the following expressions are also independent of the choice of coordinates: the stress intensity

$$\sigma_i = \frac{1}{\sqrt{2}} \sqrt{(\sigma_1 - \sigma_2)^2 + (\sigma_2 - \sigma_3)^2 + (\sigma_3 - \sigma_1)^2} \tag{3.32}$$

and the intensity of the deformation rate

$$v_{pi} = \frac{\sqrt{2}}{3} \sqrt{(v_{p1} - v_{p2})^2 + (v_{p2} - v_{p3})^2 + (v_{p3} - v_{p1})^2}. \tag{3.33}$$

In uniaxial tension, $\sigma_1 = \sigma_0$, $\sigma_2 = \sigma_3 = 0$, $v_{p1} = v_{p0}$, and $v_{p2} = v_{p3} = -\frac{1}{2}v_{p0}$) so that $\sigma_i = \sigma_0$, and $v_{pi} = v_{p0}$.

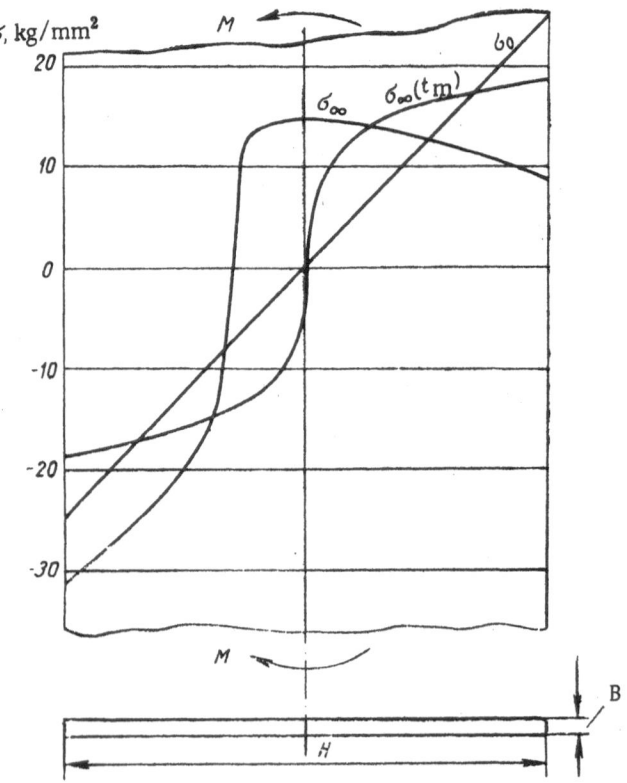

Fig. 7. Stress distribution over the thickness
of a nonuniformly heated rod in bending.

Accordingly, Eq. (2.4) (with $\Phi = 1$) may be written in the form

$$v_{pi} = Ae^{-\beta\,(T)}[e^{k\,(T)\,\sigma_i} - 1].\tag{3.34}$$

Equation (3.34) gives the fundamental relation between the creep rate intensities and the stress intensities for a compound stressed state.

Plasticity theory usually starts out with the assumption that the principal stresses are in the same direction as the deformation rates, and that the corresponding Mohr circles are similar, i.e.,

$$\frac{\sigma_1 - \sigma_2}{v_{p1} - v_{p2}} = \frac{\sigma_2 - \sigma_3}{v_{p2} - v_{p3}} = \frac{\sigma_3 - \sigma_1}{v_{p3} - v_{p1}} = c.\tag{3.35}$$

Raising the expression $\sigma_1 - \sigma_2 = c(v_{p1} - v_{p2})$, etc., to the square and adding, we find, using Eqs. (3.32) and (3.33),

$$2\sigma_i^2 = \frac{9}{2}\,c^2 v_{pi}^2,$$

from which

$$c = \frac{2}{3}\,\frac{\sigma_i}{v_{pi}}.$$

The expressions

$$\sigma_1 - \sigma_2 = \frac{2\sigma_i}{3v_{pi}}(v_{p1} - v_{p2})\quad\text{etc.}$$

may be transformed with the aid of Eqs. (3.30), (3.31), (3.34) and the condition $v_p = 0$ to the relations

$$
\left.
\begin{aligned}
v_{p1} &= \frac{3}{2}\, A e^{-\beta(T)}\, \frac{e^{k(T)\sigma_i} - 1}{\sigma_i}\, (\sigma_1 - \sigma), \\[2mm]
v_{p2} &= \frac{3}{2}\, A e^{-\beta(T)}\, \frac{e^{k(T)\sigma_i} - 1}{\sigma_i}\, (\sigma_2 - \sigma), \\[2mm]
v_{p3} &= \frac{3}{2}\, A e^{-\beta(T)}\, \frac{e^{k(T)\sigma_i} - 1}{\sigma i}\, (\sigma_3 - \sigma),
\end{aligned}
\right\}
\tag{3.36}
$$

and the reciprocal relations

$$
\left.
\begin{aligned}
\sigma_1 - \sigma &= \frac{2}{3}\, \frac{1}{k(T)\, v_{pi}}\, \ln\left[\frac{v_{pi}}{A}\, e^{\beta(T)} + 1\right] v_{p1}, \\[2mm]
\sigma_2 - \sigma &= \frac{2}{3}\, \frac{1}{k(T)\, v_{pi}}\, \ln\left[\frac{v_{pi}}{A}\, e^{\beta(T)} + 1\right] v_{p2}, \\[2mm]
\sigma_3 - \sigma &= \frac{2}{3}\, \frac{1}{k(T)\, v_{pi}}\, \ln\left[\frac{v_{pi}}{A}\, e^{\beta(T)} + 1\right] v_{p3}.
\end{aligned}
\right\}
\tag{3.37}
$$

If, instead of the principal directions 1, 2, 3 we consider the arbitrary coordinates x, y, z, Eqs. (3.36)-(3.37) continue to hold (if the subscripts 1, 2, 3 are replaced by x, y, z), and the relation between the tangential stresses σ_{xy} and the shear rate ω_{pxy} is given by the formula

$$
\omega_{p.xy} = 3 A e^{-\beta(T)}\, \frac{e^{k(T)\sigma_i} - 1}{\sigma_i}\, \sigma_{xy},
\tag{3.38}
$$

or

$$
\sigma_{xy} = \frac{1}{3}\, \frac{1}{k(T)\, v_{pi}}\, \ln\left[\frac{v_{pi}}{A}\, e^{\beta(T)} + 1\right] \omega_{p.xy}.
$$

Similar formulas give the relation between the tangential stresses σ_{yz} and σ_{zx} and the shear rates ω_{pyz} and ω_{pzx}, respectively.

Here the expressions (3.32) and (3.33) take the form

$$
\left.
\begin{aligned}
\sigma_i &= \frac{1}{\sqrt{2}}\, \sqrt{(\sigma_x - \sigma_y)^2 + (\sigma_y - \sigma_z)^2 + (\sigma_z - \sigma_x)^2 + 6(\sigma_{xy}^2 + \sigma_{yz}^2 + \sigma_{zx}^2)} \\[2mm]
v_{pi} &= \frac{\sqrt{2}}{3}\, \sqrt{(v_{px} - v_{py})^2 + (v_{py} - v_{pz})^2 + (v_{pz} - v_{px})^2 + \frac{3}{2}(\omega_{pxy}^2 + \omega_{pyz}^2 + \omega_{pzx}^2)}.
\end{aligned}
\right\}
\tag{3.39}
$$

Equations (3.36)-(3.39) establish a relation between the stresses and the stabilized creep rates in the compound stressed state. If the stresses are given, these equations uniquely determine the deformation rates. If it is partly rates, and partly stresses that are given (including at least one of the normal stresses), all the remaining rates and stresses are likewise uniquely determined. But if no one of the normal stresses is known, the relation given only makes it possible to find the differences $\sigma_x - \sigma$, etc., while the value of the mean stress remains undetermined.

Equations (3.36)-(3.39) are the "physical" conditions of the problem. In addition to these equations, the stresses must satisfy the equilibrium conditions, and the deformation rates must be related to the displacement rates through the geometric conditions, and must satisfy the boundary conditions of the problem.

133

The equilibrium conditions in rectangular coordinates are of the form

$$\left.\begin{array}{l} \dfrac{\partial \sigma_x}{\partial x} + \dfrac{\partial \sigma_{xy}}{\partial y} + \dfrac{\partial \sigma_{xz}}{\partial z} + X = 0, \\[2mm] \dfrac{\partial \sigma_y}{\partial y} + \dfrac{\partial \sigma_{yz}}{\partial z} + \dfrac{\partial \sigma_{yx}}{\partial x} + Y = 0, \\[2mm] \dfrac{\partial \sigma_z}{\partial z} + \dfrac{\partial \sigma_{zx}}{\partial x} + \dfrac{\partial \sigma_{zy}}{\partial y} + Z = 0, \end{array}\right\} \tag{3.40}$$

where X, Y, and Z are the volume forces acting in the corresponding directions. Equations (3.40) omit the inertial forces, which are negligibly small even in unstabilized creep, because of the very small rates of motion.

The boundary conditions on the surface of the body are as follows:

$$\sigma_x \cos(n,\,x) + \sigma_{xy} \cos(n,\,y) + \sigma_{xz} \cos(n,\,z) = q_x \quad \text{etc.} \tag{3.41}$$

where n, x is the angle between the normal to the surface of the body and the X axis, and q_x is the surface loading in the direction of the X axis (and similarly for the Y and Z axes).

The geometric equations are of the form:

$$\left.\begin{array}{ll} v_{pr} = \dfrac{\partial u_x}{\partial x} & {}^{(i)}_{pxy} = \dfrac{\partial u_x}{\partial y} + \dfrac{\partial u_y}{\partial x}, \\[2mm] v_{py} = \dfrac{\partial u_y}{\partial y} & \omega_{pyz} = \dfrac{\partial u_y}{\partial z} + \dfrac{\partial u_z}{\partial y}, \\[2mm] v_{pz} = \dfrac{\partial u_z}{\partial z} & \omega_{pzx} = \dfrac{\partial u_z}{\partial x} + \dfrac{\partial u_x}{\partial z}, \end{array}\right\} \tag{3.42}$$

where u_x, u_y, and u_z are the displacement rates along the X, Y, and Z axes. The displacement rates may be given on the surface (or part of the surface) of the body.

The fifteen equations (3.36), (3.38), (3.40), and (3.42), containing fifteen unknowns (6 stresses, 6 deformation rates, and 3 displacement rates), together with the boundary conditions, form the complete system of equations of stabilized creep theory.

Hollow Cylinder under a Pressure Difference and an Axial Force

Consider a hollow, thick-walled cylinder, closed at the ends, with inner radius a, outer radius b, subjected to the internal pressure p_a and the external pressure p_b, in tension under the external axial force p_0 (Fig. 8). The temperature distribution over the thickness is known: T = T(r), where r is the radius at any point.

If u_r is the radial displacement rate of a point on the radius r, and v_r and v_Θ are, respectively, the radial and peripheral deformation rates,[*] it follows from the geometric relations that

$$v_r = \frac{du_r}{dr}, \qquad v_\Theta = \frac{u_r}{r}. \tag{3.43}$$

In view of the incompressibility condition, the axial rate v_x is equal to

$$v_x = -\left(\frac{du_r}{dr} + \frac{u_r}{r}\right). \tag{3.44}$$

If the cylinder is quite long, all the cross sections far from the ends are under idential conditions, so that the rate v_x is constant

$$v_x = \text{const.} \tag{3.45}$$

---

[*] The subscript p is omitted from the creep deformation rates.

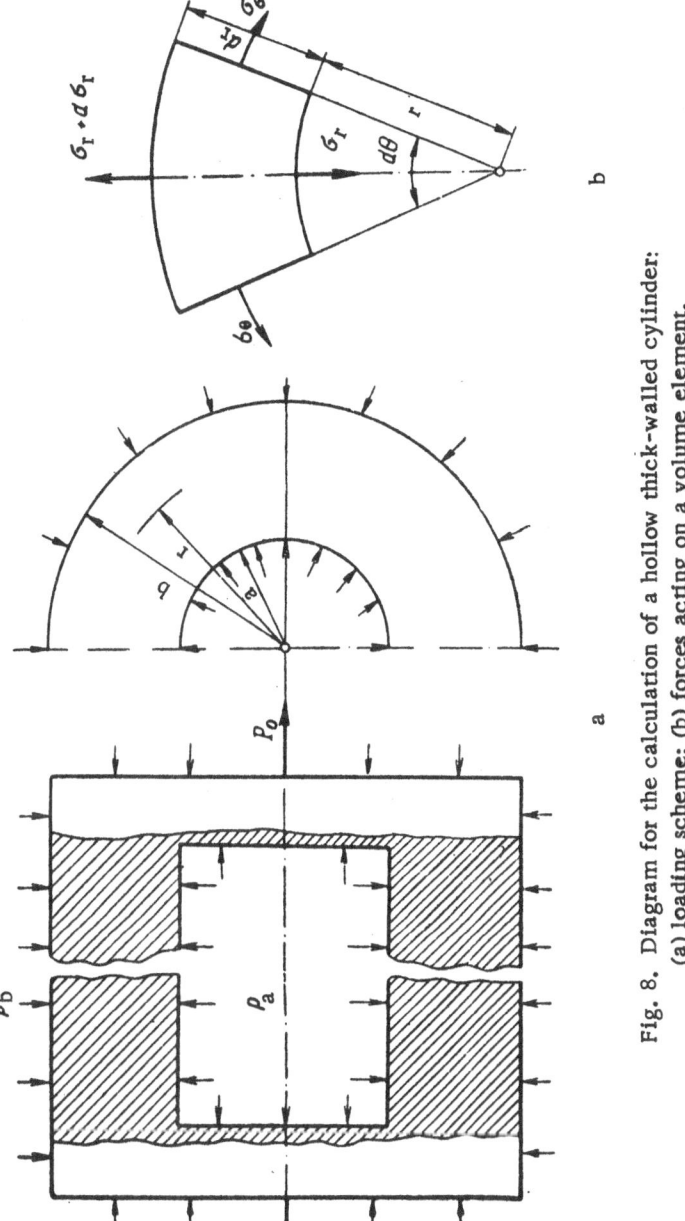

Fig. 8. Diagram for the calculation of a hollow thick-walled cylinder:
(a) loading scheme; (b) forces acting on a volume element.

Integrating Eq. (3.44) using Eq. (3.45) gives

$$u_r = \frac{c}{r} - \frac{r v_x}{2},$$

(3.46)

where c is an integration constant. We find from Eqs. (3.43) and (3.46)

$$\left. \begin{array}{l} v_r = -\dfrac{c}{r^2} - \dfrac{v_x}{2}, \\[2mm] v_\Theta = \dfrac{c}{r^2} - \dfrac{v_x}{2}. \end{array} \right\}$$

(3.47)

It follows from the symmetry conditions that the rates v_r, v_Θ, and v_x are the principal ones, so that the intensity v_i is equal to

$$v_i = \frac{\sqrt{2}}{3} \sqrt{(v_r - v_\Theta)^2 + (v_\Theta - v_x)^2 + (v_x - v_r)^2} = \sqrt{v_x^2 + \frac{4c^2}{3r^4}}.$$

(3.48)

The stress intensity from Eq. (3.34) is equal to

$$\sigma_i = \frac{1}{k(T)} \ln \left[\frac{e^{\lambda(T)}}{A} \sqrt{v_x^2 + \frac{4c^2}{3r^4}} + 1 \right].$$

(3.49)

Using the notation

$$\widetilde{G} = \frac{\sigma_i}{3 v_i} = \frac{1}{3k(T)} \left(v_x^2 + \frac{4c^2}{3r^4} \right)^{-\frac{1}{2}} \ln \left[\frac{e^{\lambda(T)}}{A} \sqrt{v_x^2 + \frac{4c^2}{3r^4}} + 1 \right],$$

(3.50)

we find from Eq. (3.37)

$$\left. \begin{array}{l} \sigma_r - \sigma = 2\widetilde{G} v_r, \\[2mm] \sigma_\Theta - \sigma = 2\widetilde{G} v_\Theta, \\[2mm] \sigma_x - \sigma = 2\widetilde{G} v_x. \end{array} \right\}$$

(3.51)

From the equation for radial equilibrium of an element of the cylinder (see Fig. 8)

$$\frac{d}{dr}(r\sigma_r) - \sigma_\Theta = 0.$$

It follows that

$$\frac{d\sigma_r}{dr} = \frac{\sigma_\Theta - \sigma_r}{r} = 2\widetilde{G} \frac{v_\Theta - v_r}{r}$$

(3.52)

and

$$\sigma_r(r) = 2 \int_a^r \widetilde{G}(r_1) \frac{v_\Theta(r_1) - v_r(r_1)}{r_1} dr_1 - p_a,$$

(3.53)

where the boundary condition $\sigma_r(a) = -p_a$ is used.

From the second condition, $\sigma_r(b) = -p_b$, we obtain

$$p_a - p_b = 2 \int_a^b \widetilde{G} \frac{v_\Theta - v_r}{r} dr = 4c \int_a^b \frac{\widetilde{G}}{r^3} dr.$$

(3.54)

Knowing σ_r, we find from Eq. (3.51)

$$\left.\begin{array}{l} \sigma_\Theta = \sigma_r + 2\widetilde{G}\,(v_\Theta - v_r), \\ \sigma_x = \sigma_r + 2\widetilde{G}\,(v_x - v_r). \end{array}\right\} \tag{3.55}$$

The axial equilibrium condition requires that

$$2\pi \int\limits_a^b \sigma_x\, r\, dr = P_0 + \pi\,(p_a a^2 - p_b b^2),$$

from which

$$\frac{P_0 + \pi\,(p_a a^2 - p_b b^2)}{2\pi} = \int\limits_a^b \sigma_r\, r\, dr + 2 \int\limits_a^b \widetilde{G} r\,(v_x - v_r)\, dr. \tag{3.56}$$

Putting into Eq. (3.56) the value of σ_r from Eq. (3.53) and integrating the first term on the right-hand side by parts, gives

$$\int\limits_a^b \sigma_r\, r\, dr = 2 \int\limits_a^b r \int\limits_a^r \widetilde{G}\,\frac{v_\Theta - v_r}{r}\, dr\, dr - p_a\,\frac{b^2 - a^2}{2}$$

$$= 2\,\frac{r^2}{2} \int\limits_a^r \widetilde{G}\,\frac{v_\Theta - v_r}{r}\, dr\,\Big|_a^b - 2 \int\limits_a^b \frac{r^2}{2}\,\widetilde{G}\,\frac{v_\Theta - v_r}{r}\, dr - p_a\,\frac{b^2 - a^2}{2}$$

$$= b^2 \int\limits_a^b \widetilde{G}\,\frac{v_\Theta - v_r}{r}\, dr - \int\limits_a^b \widetilde{G} r\,(v_\Theta - v_r)\, dr - p_a\,\frac{b^2 - a^2}{2}.$$

Using Eq. (3.54), this gives

$$\int\limits_a^b \sigma_r\, r\, dr = \frac{(p_a - p_b)\,b^2}{2} - \int\limits_a^b \widetilde{G} r\,(v_\Theta - v_r)\, dr - p_a\,\frac{b^2 - a^2}{2} = \frac{p_a a^2 - p_b b^2}{2} - \int\limits_a^b \widetilde{G} r\,(v_\Theta - v_r)\, dr.$$

Substituting the value of $\int\limits_a^b \sigma_r\, r\, dr$ in Eq. (3.56), we obtain

$$\frac{P_0}{2\pi} = \int\limits_a^b \widetilde{G} r\,(2v_x - 2v_r - v_\Theta + v_r)\, dr$$

and, finally, noting that $v = 0$,

$$\frac{P_0}{2\pi} = 3v_x \int\limits_a^b \widetilde{G}\, r\, dr. \tag{3.57}$$

From Eqs. (3.54) and (3.57), we obtain

$$c = \frac{p_a - p_b}{4\int\limits_a^b \widetilde{G}(r)\,\dfrac{dr}{r^3}}\ ,\quad v_x = \frac{p_0}{6\pi \int\limits_a^b \widetilde{G}(r)\, r\, dr}\ . \tag{3.58}$$

137

Since the function $\widetilde{G}(r)$ depends only on the constants c and v_x, Eq. (3.58) makes it possible to find the values of these constants, after which the stresses are easily found. In the general case, the constants are found by the method of successive approximations, in a way similar to that given in the section on bending calculations.

It follows from Eqs. (3.58) that:

a. at $p_a = p_b$ (compression on all sides), c = 0, and then $v_r = v_\Theta = -v_x/2$, which corresponds with the problem of the rod in tension;

b. for $P_0 = 0$ (no external axial force), $v_x = 0$, i.e., the length of the cylinder does not change, as is confirmed by experiments.

The formulas used to calculate the stresses are of the form

$$\left.\begin{aligned}
\sigma_r &= 4c \int_a^r \widetilde{G}(r_1) \frac{dr_1}{r_1^3} - p_a, \\
\sigma_\Theta &= 4c \left(\frac{\widetilde{G}}{r^2} + \int_a^r \widetilde{G}(r_1) \frac{dr_1}{r_1^3} \right) - p_a, \\
\sigma_x &= 3\widetilde{G}v_x + 2c \left(\frac{\widetilde{G}}{r^2} + 2\int_a^r \widetilde{G}(r_1) \frac{dr_1}{r_1^3} \right) - p_a.
\end{aligned}\right\} \tag{3.59}$$

Consider the special case where $P_0 = 0$ ($v_x = 0$), and $e^{k\sigma i} \gg 1$. From Eq. (3.50) we have

$$\widetilde{G} = \frac{r^2}{2\sqrt{3}\,ck(T)} \left[\ln\left(\frac{2c}{\sqrt{3}\,Ar^2} \right) + \beta(T) \right],$$

or, in somewhat different form,

$$\widetilde{G}\left(\frac{2c}{\sqrt{3}\,Aa^3} \right) = \left(\frac{r}{a} \right)^2 \frac{1}{3Ak(T)} \left[\ln\left(\frac{2c}{\sqrt{3}\,Aa^2} \right) + 2\ln\left(\frac{a}{r} \right) + \beta(T) \right]. \tag{3.60}$$

Putting $\widetilde{G}$ in the first equation of (3.58), we obtain

$$p_a - p_b = 4c \left(\frac{\sqrt{3}\,Aa^2}{2c} \right) \frac{1}{3Aa^2} \int_a^b \frac{1}{rk(T)} \left[\ln\left(\frac{2c}{\sqrt{3}\,Aa^2} \right) + 2\ln\left(\frac{a}{r} \right) + \beta(T) \right] dr,$$

from which

$$\ln\left(\frac{2c}{\sqrt{3}\,Aa^2} \right) = \frac{\dfrac{\sqrt{3}}{2}(p_a - p_b) - \displaystyle\int_a^b \frac{2\ln(a/r) + \beta(T)}{rk(T)}\,dr}{\displaystyle\int_a^b \frac{dr}{rk(T)}}. \tag{3.61}$$

Equation (3.61) may also be used to find the value of c to a first approximation in those cases where unity cannot be neglected in Eqs. (3.49) and (3.50).

Example 3. Find the steady-state stress distribution in a nonuniformly heated tube with the diameter ratio b/a = 2 under an internal pressure $p_a = 18$ atm, and find what residual stresses occur in the tube after cooling and unloading.

The temperature on the inner wall is $t_a = 750°C$, and on the outer wall $t_b = 650°C$. The material of the tube is a heat-resistant alloy for which, in the temperature range 600-800°C, the creep parameters have the following values:

$$A = 43 \ 1/h,$$

$$\beta(T) = \frac{2.2 \cdot 10^4}{T},$$

$$k(T) = 0.15 + 0.001 (T - 973).$$

Solution. The steady-state temperature field in the tube is given by the equation [14]:

$$T = T_a + (T_b - T_a) \frac{\ln r/a}{\ln b/a}.$$

Figure 9 gives the results of calculating the stresses $\sigma_r(\infty)$, $\sigma_\Theta(\infty)$, and $\sigma_x(\infty)$ from Eqs. (3.59), while Fig. 10 gives the corresponding values of the stress intensity $\sigma_i(\infty)$. The same figures show the elastic stress distribution from the internal pressure alone, $\sigma^P(0)$, and from the combined effect of pressure and nonuniform heating $\sigma(0)$, as found from the usual formulas for the strength of materials [15].

The residual stresses are found as the difference between the stabilized creep stresses and the corresponding elastic stresses. It must be kept in mind here that on unloading, the intensities σ_i cannot be subtracted; they can only be found after calculating the stress components σ_r, σ_Θ, and σ_x from Eq. (3.32).

Figures 11-13 give the residual stress distribution σ_{res} for partial unloading (the pressure is taken off, but the nonuniform heating remains in Fig. 11 and, the other way around in Fig. 12), and for complete unloading (Fig. 13).

Calculation shows that creep produces a considerable stress redistribution during operation, and leaves large residual stresses after unloading and cooling the part.

Deformation of a Rod in the Form of a Solid Circular Cylinder

A discussion has already been given of stabilized creep in a nonuniformly heated rod of arbitrary cross section under an axial force, where it was assumed, in accordance with the theory of rods, that the material is in uniaxial tension.

The problem may be solved exactly for a rod of circular cross section.

From the condition that at the center of the cylinder, at $r = 0$, the radial velocity u_r does not go to infinity. Using Eqs. (3.43)-(3.46), it follows that $c = 0$, and

$$\left. \begin{aligned} u_r &= -r \frac{v_x}{2}, \\ v_r = v_\Theta &= -\frac{v_x}{2} = \text{const}, \\ v_i &= |v_x| = \text{const}. \end{aligned} \right\} \tag{3.62}$$

The stress intensity is equal to

$$\sigma_i = \frac{1}{k(T)} \ln [ae^{\beta(T)} + 1], \tag{3.63}$$

where

$$a = \frac{1}{A} |v_x|,$$

while it follows from Eqs. (3.50)-(3.51) that over the whole volume

$$\sigma_r = \sigma_\Theta,$$

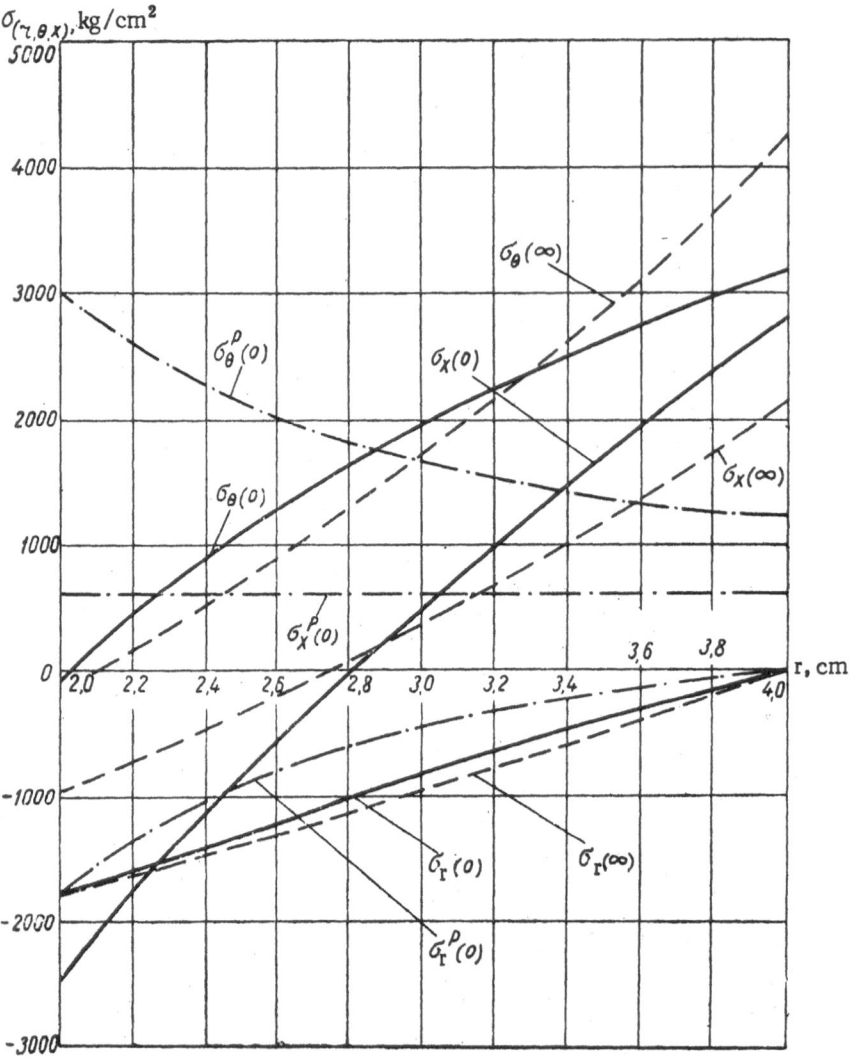

Fig. 9. Stress distribution over thickness of tube.

and, since $\sigma_r = 0$ at $r = b$, from Eq. (3.52)

$$\sigma_r \equiv \sigma_\theta \equiv 0, \qquad (3.64)$$

i.e., the rod is actually in a uniaxial stressed state.

The stress in the axial direction is equal to

$$\sigma_x = \frac{\operatorname{sign} v_x}{k\,(T)} \ln\left(ae^{\beta/T} + 1\right). \qquad (3.65)$$

The constant a is found from the equilibrium condition

$$2\pi \int_0^b \sigma_x\, r\, dr = P_0. \qquad (3.66)$$

The results of the calculation are in complete agreement with the data of the first paragraph of Section 3.

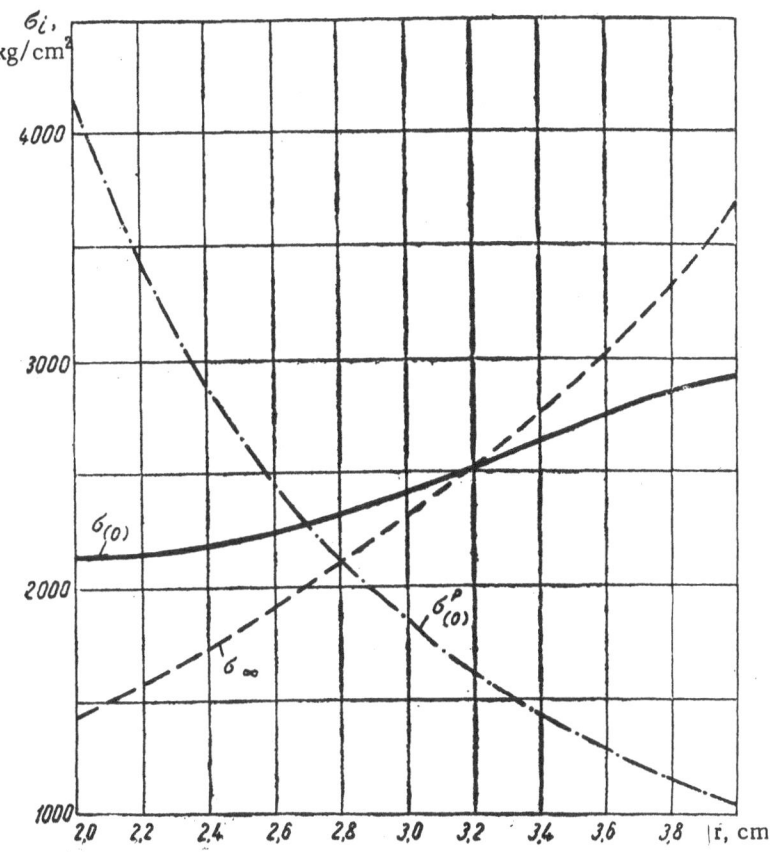

Fig. 10. Distribution of stress intensity over thickness of tube.

Creep of a Plate

Consider a rectangular plate of thickness h, in uniform tension, and bent in the two directions x and y (Fig. 14) in such a way that at the cross sections x = const the longitudinal force N_x and the bending moment M_y are acting (per unit length) and, similarly, the force N_y and the moment M_x are acting at the cross sections y = const. It is obvious that

$$
\left.
\begin{aligned}
N_x &= \int_{-h/2}^{+h/2} \sigma_x \, dz, & N_y &= \int_{-h/2}^{+h/2} \sigma_y \, dz, \\
M_y &= -\int_{-h/2}^{+h/2} \sigma_x \, z dz, & M_x &= \int_{-h/2}^{+h/2} \sigma_y \, z dz,
\end{aligned}
\right\}
\tag{3.67}
$$

where σ_x and σ_y are the normal stresses in the cross sections x, y, and the z coordinate is read from the median surface of the plate ($-h/2 \le z \le +h/2$).

We limit ourselves to the simplest case, where the tension and bending are the same in both directions, i.e., $N_x = N_y = N$, $M_x = M_y = M$, and $\sigma_x = \sigma_y$. Since the plate is thin, we may assume $\sigma_z \approx 0$, i.e.,

$$
\sigma_i = \frac{1}{\sqrt{2}} \sqrt{(\sigma_x - \sigma_y)^2 + \sigma_x^2 + \sigma_y^2} = \sqrt{\sigma_x^2 - \sigma_x \sigma_y + \sigma_y^2},
\tag{3.68}
$$

which, for the condition $\sigma_x = \sigma_y$, gives $\sigma_i = |\sigma|$.

141

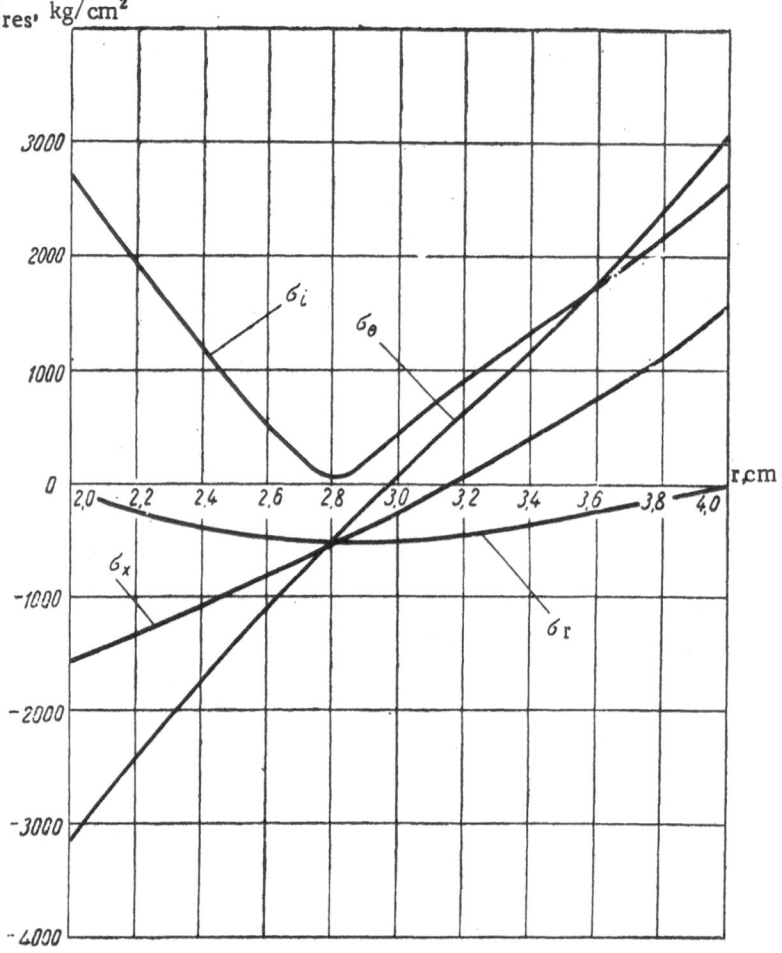

Fig. 11. Residual stress distribution over thickness of tube
with partial unloading – internal pressure removed.

The cross sections of the plate remain plane in uniform tension and bending, i.e.,

$$
\left.
\begin{aligned}
\varepsilon_x &= \varepsilon_{x_0} - \varkappa_x z, \\
\varepsilon_y &= \varepsilon_{y_0} - \varkappa_y z,
\end{aligned}
\right\}
\tag{3.69}
$$

where ε_{x0} and ε_{y0} are the elongations, $\varkappa_x$ and $\varkappa_y$ are the elastic curvatures of the median plane.

On the other hand (for $\varepsilon_{p0} = 0$),

$$
\left.
\begin{aligned}
\varepsilon_x &= \frac{1}{E}\left(\sigma_x - \mu\sigma_y\right) + \varepsilon_{xp} + \dot{\alpha}t, \\
\varepsilon_y &= \frac{1}{E}\left(\sigma_y - \mu\sigma_x\right) + \varepsilon_{yp} + \alpha t.
\end{aligned}
\right\}
\tag{3.70}
$$

Setting Eq. (3.69) equal to Eq. (3.70), and differentiating with respect to the time, we obtain for the stabilized creep

$$
\left.
\begin{aligned}
v_{px} &= \frac{d\varepsilon_{x_0}}{d\tau} - \frac{d\varkappa_x}{d\tau}\, z, \\
v_{py} &= \frac{d\varepsilon_{y_0}}{d\tau} - \frac{d\varkappa_y}{d\tau}\, z,
\end{aligned}
\right\}
\tag{3.71}
$$

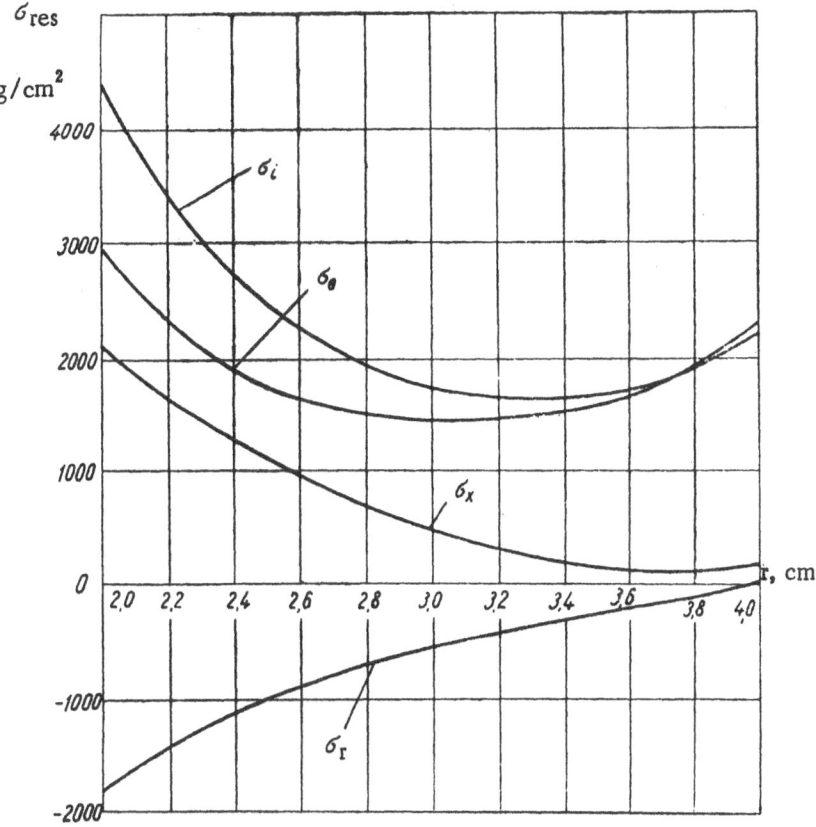

Fig. 12. Residual stress distribution over thickness of tube
with partial unloading — nonuniform heating removed.

Using the incompressibility condition, we find from Eq. (3.33) the expression for the creep rate intensity

$$v_{pi} = \frac{2}{\sqrt{3}} \sqrt{v_{px}^2 + v_{px}v_{py} + v_{py}^2},$$

and for the same conditions in the x and y directions ($v_{px} = v_{py} = v_p$)

$$v_{pi} = 2|v_p| = 2\left|\frac{d\varepsilon_0}{d\tau} - \frac{d\varkappa}{d\tau}z\right|. \tag{3.72}$$

Putting the value of v_{pi} into Eqs. (3.37) gives

$$\sigma_x = \sigma_y = \frac{1}{k(T)}\operatorname{sign}(a+bz)\ln[\,|a+bz|\,e^{s(T)} + 1], \tag{3.73}$$

where

$$a = \frac{2}{A}\frac{d\varepsilon_0}{d\tau}, \quad b = -\frac{2}{A}\frac{d\varkappa}{d\tau}.$$

The constants a and b are found from the equilibrium conditions (3.67). A comparison of the solutions for
a plate (3.73) with those for a rod (3.19) shows that for the same deformations in the directions of the X and Y
axes, the stress distribution in stabilized creep turns out to be the same over the thickness of either the plate or
the rod, but the deformation rate for the plate is a factor of two less than for the rod.

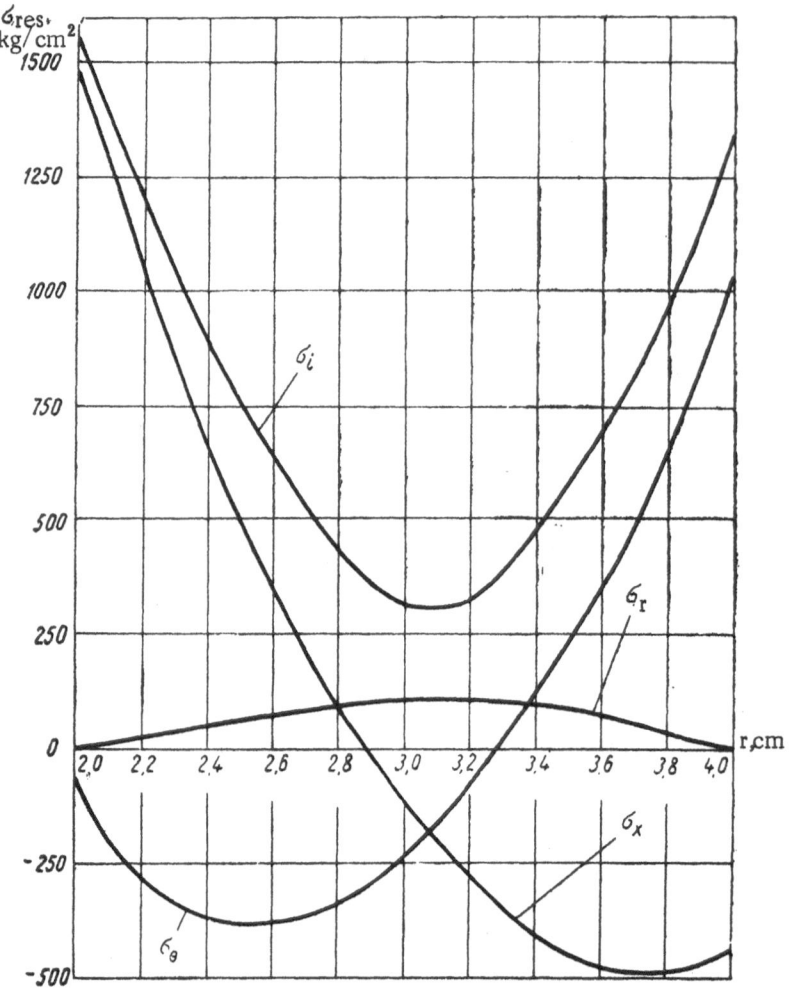

Fig. 13. Residual stress distribution over thickness of tube
with complete unloading.

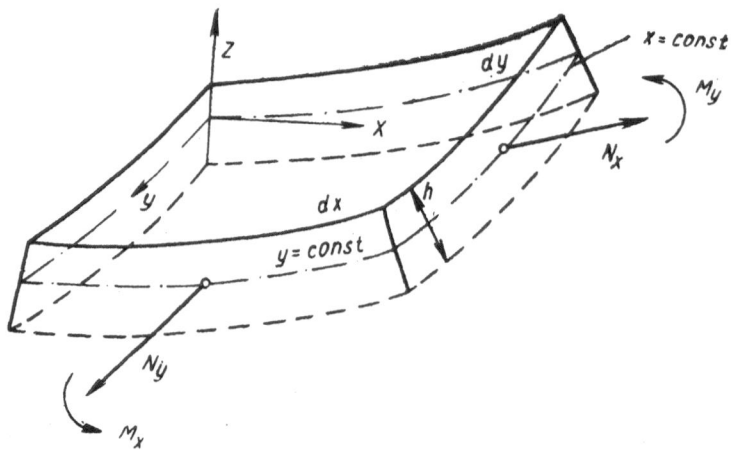

Fig. 14. Loading scheme of rectangular plate.

144

4. Unstabilized Creep of Nonuniformly Heated Parts

In accordance with the general theory, presented in Section 2, and the creep equations in a compound stressed state discussed in Section 3, the basic parameters of unstabilized creep are given by the following equations.

Creep rate intensity:

$$v_{pi} = Ae^{-\beta(T)}[e^{k(T)\sigma_i} - 1]\left(\frac{p}{p_*}\right)^{-\alpha}, \tag{4.1}$$

where

$$\alpha = \begin{cases} \alpha(T, \sigma_i) & \text{for} \quad p/p_* \leqslant 1, \\ 0 & \text{for} \quad p/p_* > 1. \end{cases} \tag{4.2}$$

The boundary of the hardening stage

$$p_* = A_* e^{-\beta_*(T)}[e^{k_*(T)\sigma_i} - 1]. \tag{4.3}$$

Accumulated plastic deformation:

$$p(\tau) = \int_0^\tau v_{pi}(\tau_1)\,d\tau_1. \tag{4.4}$$

The components of the instantaneous plastic deformation:

$$\left.\begin{aligned}
\varepsilon_x^p(\tau) &= \int_0^\tau v_{px}(\tau_1)\,d\tau_1 = \frac{3}{2}\int_0^\tau \frac{v_{pi}(\tau_1)}{\sigma_i(\tau_1)}[\sigma_x(\tau_1) - \sigma(\tau_1)]\,d\tau_1, \\
&\cdots\cdots\cdots\cdots\cdots\cdots\cdots\cdots\cdots \\
\gamma_{xy}^p(\tau) &= \int_0^\tau \omega_{pxy}(\tau_1)\,d\tau_1 = 3\int_0^\tau \frac{v_{pi}(\tau_1)}{\sigma_i(\tau_1)}\sigma_{xy}(\tau_1)\,d\tau_1, \\
&\cdots\cdots\cdots\cdots\cdots\cdots\cdots\cdots\cdots
\end{aligned}\right\} \tag{4.5}$$

etc. The intensities v_{pi} and σ_i are found from Eqs. (3.39).

It is perfectly obvious that it is impossible to solve the system of equations (4.1)-(4.5) simultaneously with Eqs. (3.40)-(3.42) in the general case, although it can be done in some of the simplest cases. However, successive calculation of the creep makes it possible in a number of problems to get a sufficiently accurate approximate solution. The way such a calculation is made will be shown in detail below using the deformation of a rod without bending as an example.

Deformation of a Rod without Bending – Successive Calculation of Creep

We return to the problem of a rod in tension (Section 3). It follows from the deformation equations (3.2) that the stress σ at any instant of time is equal to

$$\sigma = E(\varepsilon - \varepsilon_{p0} - \varepsilon_p - \alpha t), \tag{4.6}$$

where ε is a constant quantity for a given cross section.

Putting Eq. (4.6) into the equilibrium condition (3.6) gives

$$N = \int_F E(\varepsilon - \varepsilon_{p0} - \varepsilon_p - \alpha t)\,dF,$$

from which

$$\varepsilon = \frac{1}{E_m F}\left(N + \int_F E\varepsilon_{p0}\,dF + \int_F E\varepsilon_p\,dF + \int_F E\alpha t\,dF\right), \tag{4.7}$$

145

where

$$E_m = \frac{1}{F} \int\limits_F E \, dF.$$

Using (4.7), Eq. (4.6) reduces to the form

$$\sigma = \sigma_N + L(E\varepsilon_{p0}) + L(E\varepsilon_p) + L(E\alpha t), \qquad (4.8)$$

where $\sigma_N = \bar{E}\sigma_m$ $(\bar{E} = E/E_m,\ \sigma_m = N/F)$ is the stress from the external force, and the notation used is

$$L(f) = \frac{\bar{E}}{F} \int\limits_F f \, dF - f, \qquad (4.9)$$

the function f being understood to be any of the deformations: ε_{p0}, ε_p, or αt.

If there is no creep, we have

$$\sigma = \sigma^\circ = \sigma_N + L(E\varepsilon_{p0}) + L(E\alpha t). \qquad (4.10)$$

The purely elastic stresses (for $\varepsilon_{p0} = 0$) are calculated directly from Eqs. (4.10). When the stresses exceed the elastic limit ($\varepsilon_{p0} \neq 0$), the stresses are found from the deformation diagrams of the material, which give $\sigma = f(\varepsilon_{p0})$ at the appropriate temperature. In a creep calculation, the stress σ^0 may be regarded as fixed, while in the general case it changes with time (but this change will not be too rapid for the problem to be regarded as quasi-static).

It follows from Eqs. (4.8) and (4.10) that

$$\sigma = \sigma^\circ + L(E\varepsilon_p). \qquad (4.11)$$

Equation (4.11) makes it possible to find the distribution of σ, if the distribution of ε_p is known. In particular, for $E = \mathrm{const}$, we have

$$L(E\varepsilon_p) = E(\varepsilon_{p\,m} - \varepsilon_p).$$

A second equation relating σ with ε_p is obtained from the creep equations.

Assume that at the instant of time τ_n, and at every point in the cross section, we know the stresses σ_n, the creep deformations ε_{pn}, and the accumulated plastic deformation p_n. In the uniaxial stressed state, $\sigma_i = |\sigma|$, and $v_{pi} = |v_p|$, so that if follows from Eq. (4.4) that:

$$p(\tau) = \int\limits_0^\tau |v_p(\tau_1)| \, d\tau_1 \text{ and } \frac{dp}{d\tau} = |v_p| \qquad (4.12)$$

(while $d\varepsilon_p/d\tau = v_p$).

Equation (4.1) takes the form

$$p^\alpha \frac{dp}{d\tau} = Ae^{-\beta(T)}[e^{k(T)|\sigma|} - 1] p_*^\alpha (T, \sigma). \qquad (4.13)$$

In order to get to the values of σ_{n+1}, $\varepsilon_{p,n+1}$, and p_{n+1} at the next instant of time τ_{n+1}, separated from τ_n by the time interval $\triangle\tau_n = \tau_{n+1} - \tau_n$, we integrate Eq. (4.13) from τ_n to τ_{n+1}

$$\chi(p_{n+1}) = \chi(p_n) + A \int\limits_{\tau_n}^{\tau_{n+1}} e^{-\beta(T)}[e^{k(T)|\sigma|} - 1] p_*^\alpha (T, \sigma) \, d\tau, \qquad (4.14)$$

where

$$\chi(p) = \int_0^p p_1^\alpha \, dp_1.$$

If α does not change with time, we have

$$\chi(p) = \frac{p^{\alpha+1}}{\alpha+1}. \tag{4.15}$$

Both the temperature T and the stress σ can change with time in an arbitrary way. However, if the time interval $\Delta\tau_n$ is quite small, it may be assumed that during this time $d\chi/d\tau$ remains constant, so that

$$\chi(p_{n+1}) = \chi(p_n) + Ae^{-\beta(T_n)}[e^{k(T_n)|\sigma_n|} - 1]\, p_{*n}^{\alpha_n} \Delta\tau_n, \tag{4.16}$$

where

$$T_n = T(\tau_n), \quad \sigma_n = \sigma(\tau_n) \text{ и } p_{*n}^{\alpha_n} = p_*^{\alpha(\tau_n)}(\tau_n).$$

To find the permissible value of $\Delta\tau_n$, consider the case of very large stresses, where Eqs. (2.11)-(2.12) hold. Replace the actual curves $\beta_0[T(\tau)]$ and $k_0|\sigma(\tau)|$ by segments of straight lines in the interval from τ_n to τ_{n+1}; thus

$$\left.\begin{aligned}
\beta_0(\tau_1) &= \beta_{0,n} + \Delta\beta_{0,n}\frac{\tau_1 - \tau_n}{\Delta\tau_n}, \\
k_0|\sigma(\tau_1)| &= k_0\sigma_n + \Delta k_0|\sigma|_n\frac{\tau_1 - \tau_n}{\Delta\tau_n},
\end{aligned}\right\} \tag{4.17}$$

where

$$\tau_n \leqslant \tau_1 \leqslant \tau_{n+1}; \quad \Delta\beta_{0,n} = \beta_{0,n+1} - \beta_{0,n};$$
$$\Delta k_0|\sigma|_n = k_{0,n+1}|\sigma_{n+1}| - k_{0,n}|\sigma_n|.$$

Substituting Eqs. (4.17) in (4.14), using Eqs. (2.11)-(2.12), and integrating, we obtain

$$\chi(p_{n+1}) = \chi(p_n) + A_0 e^{-\beta_{0,n}} \left[e^{k_0|\sigma|_n} \frac{e^{\Delta k_0|\sigma|_n - \Delta\beta_{0n}} - 1}{\Delta k_0|\sigma|_n - \Delta\beta_{0,n}} - \frac{1 - e^{-\Delta\beta_{0,n}}}{\Delta\beta_{0,n}} \right]\Delta\tau_n. \tag{4.18}$$

If the time interval $\Delta\tau_n$ is taken sufficiently small to satisfy the condition

$$\frac{1}{2}|\Delta k_0|\sigma|_n - \Delta\beta_{0,n}|_{max} \leqslant \delta^2, \tag{4.19}$$

where $\delta^2 \ll 1$, then (from the expansion $e^x = 1 + x + x^2/2 + \ldots$) from which it follows that $(e^x-1)/x = 1 + x/2 + \ldots = 1$ at $x/2 \ll 1$ we obtain

$$\chi(p_{n+1}) = \chi(p_n) + A_0 e^{-\beta_{0,n}} \left[e^{k_0|\sigma|_n} - \frac{1 - e^{-\Delta\beta_{0,n}}}{\Delta\beta_{0,n}} \right]\Delta\tau_n, \tag{4.20}$$

and, finally, at the steady-state temperature

$$\chi(p_{n+1}) = \chi(p_n) + A_0 e^{-\beta_0}(e^{k_0|\sigma|_n} - 1)\Delta\tau_n. \tag{4.21}$$

Equation (4.19) serves to find the allowable value of $\Delta\tau_n$ even in the more general case of using Eq. (4.16).

Equations (4.16)-(4.21) make it possible to find the value of p_{n+1}, if we know p_n, σ_n, and the temperature change with time $T(\tau)$, and the time interval $\Delta\tau_n$ is fixed. Assuming further that during the time $\Delta\tau_n$ the sign of the stress (and this means of the deformation rate) does not change, we find from Eq. (4.12):

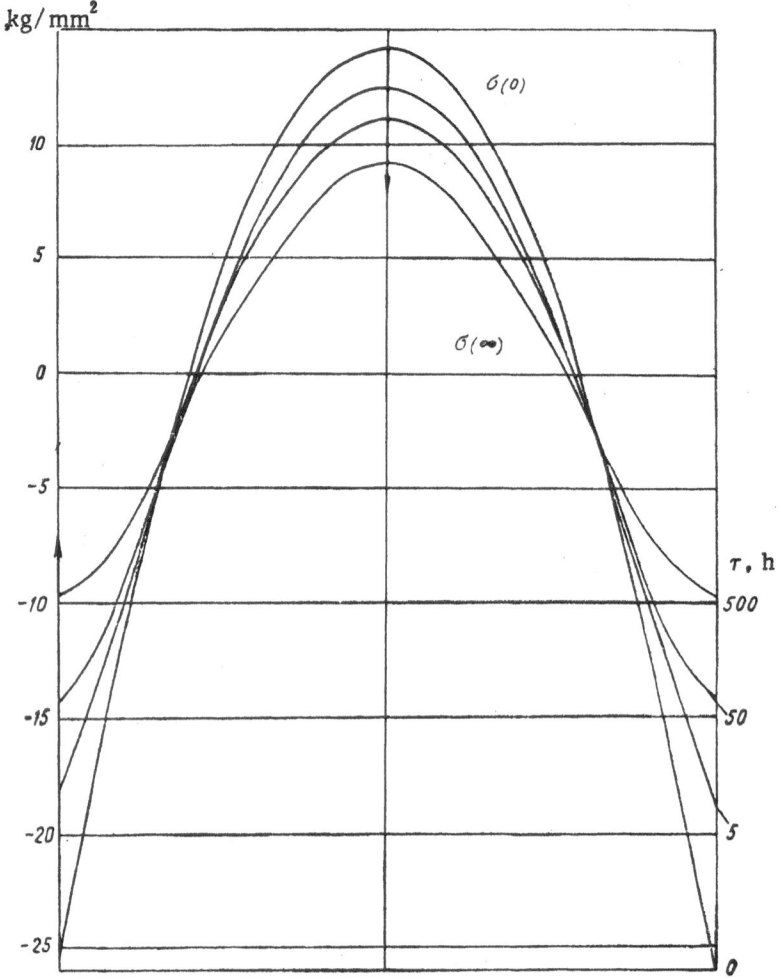

Fig. 15. Stress distribution over the thickness of a nonuniformly heated rod at different instants of time. The numbers on the right give the time in hours.

$$\varepsilon_{p,\,n+1} = \varepsilon_{p,\,n} + \text{sign}\, \sigma_n\,(p_{n+1} - p_n).\tag{4.22}$$

Equation (4.22) may also be used approximately if the sign of the stress changes during the time $\Delta \tau_n$, since, in this case, creep is practically absent, so that $\varepsilon_{p,n+1} \approx \varepsilon_{p,n}$.

Equations (4.16)-(4.22), together with Eq. (4.11), which in the present case is of the form

$$\sigma_{n+1} = \sigma_{n+1}^0 + L\,(E\varepsilon_{p,\,n+1}),\tag{4.23}$$

permit making a creep calculation for successive time in a rod in tension (in compression) in a nonuniform transient temperature field under varying load. The stress σ_0 is known at the initial instant of time — it is found from Eq. (4.10), while the deformations ε_{p0} and p_0 are equal to zero (if the part has not been subjected to any previous plastic deformation). With the value of $\Delta \tau_0$ fixed, we find the value of p_1 from Eq. (4.18) or (4.20), then ε_{p1} from Eq. (4.22), and the value of σ_1 from Eq. (4.23). If the value of $k_0\,|\sigma|_1 = k_{0,1}|\,\sigma_1\,| - k_{0,0}\,|\sigma_0|$ satisfies condition (4.19), we pass to the next stage of the calculation, but if this condition is not satisfied, we have to reduce the value of $\Delta \tau_0$ and do the calculation over.

The allowable value of $\Delta \tau_0$ for the first time interval with T = const and N = const may be found analytically. We have from Eqs. (4.21), (4.15), and (4.22):

$$\frac{p_1^{\alpha+1}}{\alpha+1} = A_0 e^{-\beta_0(T)}\,[e^{k_0\,|\,\sigma_0\,|} - 1]\,\Delta\tau_0,\tag{4.24}$$

148

$$\varepsilon_{|p_1|} = \text{sign } \sigma_0 p_1 = \text{sign } \sigma_0 \left(\frac{A_0}{m} \right)^m e^{-m\beta_0(T)} [e^{k_0 | \sigma_0 |} - 1]^m \Delta \tau_0^m, \tag{4.25}$$

where m = 1/(α + 1).

Substituting Eq. (4.25) in (4.23), we find

$$\sigma_1 = \sigma_0 + L(E\varepsilon_{p_1}) = \sigma_0 + L(q) \Delta \tau_0^m, \tag{4.26}$$

where q is a known function depending on the initial stresses σ_0:

$$q = \left(\frac{A_0}{m} \right)^m E \text{ sign } \sigma_0 e^{-m\beta_0(T)} [e^{k_0 | \sigma_0 |} - 1]^m. \tag{4.27}$$

Assuming that the sign of σ does not change during the time $\Delta \tau_0$ (at any rate at the points where the stresses σ_0 are large), we obtain from Eq. (4.19)

$$\frac{1}{2} k_0 | \Delta | \sigma |_0 | = \frac{1}{2} k_0 | | \sigma_1 | - | \sigma_0 | | = \frac{1}{2} k_0 | \sigma_1 - \sigma_0 | = \frac{1}{2} k_0 | L(q) | \Delta \tau_0^m \leqslant \delta^2, \tag{4.28}$$

and, leaving, for the points where $|L(q)|$ is a maximum, the equality sign alone, we have

$$\Delta \tau_0 = \left[\frac{2\delta^2}{k_0 | L(q) |_{\max}} \right]^{1/m}. \tag{4.29}$$

Since the creep rate drops off sharply with time, successively increasing values may, as a rule, be taken for the succeeding time intervals [usually, $\Delta \tau_{n+1} = (3-5)\Delta \tau_n$].

As in stabilized creep, the condition ε = const is identically satisfied only in tension of a rod with a doubly symmetric cross section and the corresponding temperature field. In the general case, the coordinates of the longitudinal rigidity pole are given by the equations

$$y_i = \frac{1}{N} \int_F \sigma y \, dF, \quad z_i = \frac{1}{N} \int_F \sigma z \, dF, \tag{4.30}$$

and since σ changes with time, the coordinates y_i and z_i are variable.

Example 4. Using the conditions of example 1, find the stress distribution in a rod in tension at different instants of time for the following forms of the problem:

a. Longitudinal force N = 2.5 tons, thermal stresses absent (take α = 0 for this case),

b. Thermal stresses at N = 0 (magnitude α = 2 · 10^{-5} 1/°C),

c. The longitudinal force N = 2.5 tons and the temperature stresses acting together.

For simplicity, we limit ourselves to the range p < p_*, and assume p_* = const, so that $A_0 = A_{p_*}^\alpha$, $\beta_0 = \beta$, and $k_0 = k$.

Figure 4 gives the results of a calculation for "a" and "c," while Fig. 15 gives the results for "b," from which it may be seen that unstabilized creep produces a considerable change in the stress distribution, particularly in the first few hours of operation. With time, the curves for case "a" and "c" approach, from opposite sides, the limiting curves σ(∞) from example 1, and the temperature stresses gradually relax, σ(∞) → 0.

Example 5. Compare the results of exact and approximate stress relaxation calculations in a rod made of an alloy, the data for which were given in examples 1 and 4, at the constant temperature T = 780°C, and the initial stress σ_0 = 25 kg/mm².

Exact solution of the stress relaxation problem for t = const and p ⩾ p$_*$. If the total deformation ε = const, and, in particular, for the initial instant of time, $\varepsilon = \sigma_0/E$, for any instant of time, $\varepsilon = \sigma/E + \varepsilon_p$ = const, so that

$$\varepsilon_p = \frac{1}{E}\,(\sigma_0 - \sigma)\ \text{and}\ \frac{1}{E}\,\frac{d\sigma}{d\tau} = -\,\frac{d\varepsilon_p}{d\tau}\,. \tag{4.31}$$

Remembering that in the present case p = ε_p, and substituting Eq. (4.31) in Eq. (4.13), we find.

$$-\frac{1}{E^\alpha}\,(\sigma_0 - \sigma)^\alpha\,\frac{1}{E}\,\frac{d\sigma}{d\tau} = A_0 e^{-\beta_0\,(T)}\,[e^{k_0\,(T)\,\sigma} - 1], \tag{4.32}$$

from which

$$\tau = \frac{e^{\beta_0\,(T)}}{A_0 E^{\alpha+1}} \int\limits_\sigma^{\sigma_0} \frac{(\sigma_0 - \sigma_1)^\alpha}{e^{k_0 \sigma_1} - 1}\,d\sigma_1. \tag{4.33}$$

After making a numerical evaluation of the integral in Eq. (4.33), we find the function $\sigma = f\,(\tau)$.

Approximate solution. In this case, L(f) = 0, so that $\sigma = \sigma_N = \sigma_m$. From Eq. (4.6) we again find $\sigma = \sigma_0 - E\varepsilon_p$, and, in particular, for the nth approximation

$$\sigma_n = \sigma_0 - E\varepsilon_{pn}. \tag{4.34}$$

Since now $\Delta|\sigma|_0 = \sigma_0 - \sigma_1 = E\varepsilon_{p1} = q\Delta\tau_0^m$, we obtain from the condition (4.19)

$$\frac{1}{2}\,k_0 q_0\,\Delta\tau_0^m \leqslant \delta^2,$$

from which, using Eq. (4.25),

$$\Delta\tau_0 = \left(\frac{2\delta^2}{k_0 E}\right)^{1/m} \frac{me^{\beta_0\,(T)}}{A_0\,(e^{k_0\sigma_0} - 1)}\,. \tag{4.35}$$

Knowing σ_0, by the method given above, we find the values of ε_{p1}, etc., for ε_{p0} = 0.

A comparison of the results of calculations for δ^2 = 0.1 is given in Fig. 16, from which it may be seen that using Eq. (4.21) gives entirely satisfactory accuracy.

Deformation of a Rod of Arbitrary Cross Section with Bending

Substituting in the deformation equation (3.2) the expression for ε corresponding with the plane-cross-section hypothesis (3.13), we obtain

$$\sigma = E\,(\varepsilon_0 - \varkappa_y y - \varkappa_z z - \varepsilon_{p_0} - \varepsilon_p - \alpha t). \tag{4.36}$$

Substituting Eq. (4.36) in the equilibrium condition for the rod (3.17), we find:

$$\left.\begin{aligned}
\varepsilon_0 \int\limits_F E\,dF - \varkappa_y \int\limits_F Ey\,dF - \varkappa_z \int\limits_F Ez\,dF - \int\limits_F E\varepsilon_{p0}\,dF - \int\limits_F E\varepsilon_p\,dF - \int\limits_F E\alpha t\,dF &= N, \\[2mm]
\varepsilon_0 \int\limits_F Ey\,dF - \varkappa_y \int\limits_F Ey^2\,dF - \varkappa_z \int\limits_F Eyz\,dF - \int\limits_F E\varepsilon_{p0}\,y\,dF - \int\limits_F E\varepsilon_p\,y\,dF - \int\limits_F E\alpha ty\,dF &= -M_z, \\[2mm]
\varepsilon_0 \int\limits_F Ez\,dF - \varkappa_y \int\limits_F Eyz\,dF - \varkappa_z \int\limits_F Ez^2\,dF - \int\limits_F E\varepsilon_{p0}\,z\,dF - \int\limits_F E\varepsilon_p\,z\,dF - \int\limits_F E\alpha t\,z\,dF &= M_y,
\end{aligned}\right\} \tag{4.37}$$

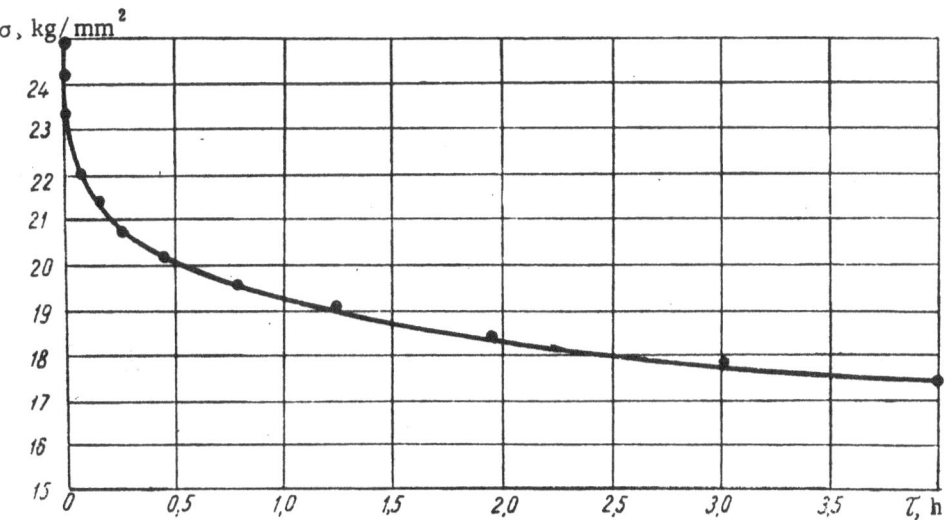

Fig. 16. Stress relaxation in a rod in tension.
Solid curve — exact solution; points — approximate solution.

Assume that the Y and Z axes are the principal reduced axes of inertia of the cross section. The direction of the axes, and the position of the origin of coordinates are found from the conditions

$$\int_F EydF = \int_F EzdF = \int_F EyzdF = 0. \tag{4.38}$$

Then it follows from Eqs. (4.37) that

$$\left.\begin{aligned}
\varepsilon_0 &= \frac{1}{E_m F}\left(N + \int_F E\varepsilon_{p0}\,dF + \int_F E\varepsilon_p\,dF + \int_F E\alpha t\,dF \right), \\
\varkappa_y &= \frac{1}{E_m I_z}\left(M_z - \int_F E\varepsilon_{p0}\,ydF - \int_F E\varepsilon_p\,ydF - \int_F E\alpha ty\,dF \right), \\
\varkappa_z &= -\frac{1}{E_m I_y}\left(M_y + \int_F E\varepsilon_{p0}\,zdF + \int_F E\varepsilon_p\,zdF + \int_F E\alpha tz\,dF \right),
\end{aligned}\right\} \tag{4.39}$$

where

$$I_y = \int_F \overline{E}z^2\,dF, \quad I_z = \int_F \overline{E}y^2\,dF. \tag{4.40}$$

Substituting Eqs. (4.39) in Eq. (4.36), we obtain the general formula giving the relation between the stress and the creep deformation:

$$\sigma = \sigma^0 + L(E\varepsilon_p), \tag{4.41}$$

where σ^0 is the stress in the rod with no creep

$$\sigma^0 = \overline{E}\left(\frac{N}{F} - \frac{M_z y}{I_z} + \frac{M_y z}{I_y} + L\left(E\varepsilon_{p0}\right) + L\left(E\alpha t\right) \right), \tag{4.42}$$

while the operator $L(f)$ is understood to be the following integral function of f:

$$L(f) = \overline{E}\left(\frac{1}{F}\int_F f\,dF + \frac{y}{I_z}\int_F fy\,dF + \frac{z}{I_y}\int_F fz\,dF \right) - f. \tag{4.43}$$

151

In order to find the position of the principal reduced axes for an arbitrary cross section, we proceed in the following way. In the arbitrary coordinate system Y_1, Z_1 we find the coordinates of the reduced center of gravity (Fig. 17)

$$y_{10} = \frac{1}{F} \int_F \overline{E} y_1 \, dF, \\ z_{10} = \frac{1}{F} \int_F \overline{E} z_1 \, dF, \qquad (4.44)$$

then calculate the reduced moments of inertia about the central axes Y_2, Z_2:

$$I_{y_2} = \int_F \overline{E} z_2^2 \, dF, \quad I_{z_2} = \int_F \overline{E} y_2^2 \, dF, \quad I_{y_2 z_2} = \int_F \overline{E} y_2 z_2 \, dF$$

and find the angle β between the axis Y_2 and the principal axis Y:

$$\tan 2\beta = \frac{2 I_{y_2 z_2}}{I_{z_2} - I_{y_2}} . \qquad (4.45)$$

Then

$$I_y = I_{y_2} \cos^2 \beta + I_{z_2} \sin^2 \beta - I_{y_2 z_2} \sin 2\beta, \\ I_z = I_{y_2} \sin^2 \beta + I_{z_2} \cos^2 \beta + I_{y_2 z_2} \sin 2\beta. \qquad (4.46)$$

The law of creep in bending retains the same form as in tension, so that the methods of calculation given in the preceding section also hold for the present problem.

Unstabilized Creep in a Compound Stressed State

The instantaneous deformation of a body in a given direction is given by Eq. (3.2) as

$$\varepsilon_x = \varepsilon_x^s + \varepsilon_x^p + \alpha t, \\ \gamma_{xy} = \gamma_{xy}^s + \gamma_{xy}^p , \qquad (4.47)$$

where ε_x^s and γ_{xy}^s are the deformations associated with the stress, as given by the instantaneous loading curve.

Assume for simplicity that the quantities ε_x^s and γ_{xy}^s are purely elastic, although the "rapid" plastic deformation can also be taken account of in principle.

Represent the total deformation as the sum of the volume deformation:

$$\theta = \varepsilon_x + \varepsilon_y + \varepsilon_z = \varepsilon_x^s + \varepsilon_y^s + \varepsilon_z^s + 3\alpha t = \theta_s + 3\alpha t = 3\varepsilon, \qquad (4.48)$$

where $\varepsilon = \varepsilon_s$ and $+\alpha t$ are the mean deformation and the shear deformation with the components

$$\varepsilon_x' = \varepsilon_x - \varepsilon, \quad \varepsilon_y' = \varepsilon_y - \varepsilon, \quad \varepsilon_z' = \varepsilon_z - \varepsilon.$$

It is assumed in Eq. (4.48) that the volume does not change in creep.

It has been found experimentally that independently of the amount of plastic deformation, the volume change is practically proportional to the mean pressure, i.e.,

$$\theta_s = \theta - 3\alpha t = \frac{\sigma}{K} , \qquad (4.49)$$

where K is the bulk modulus of elasticity $K = E/3(1 - 2\mu)$ $= 2(1 + \mu)G/3(1 - 2\mu)$.

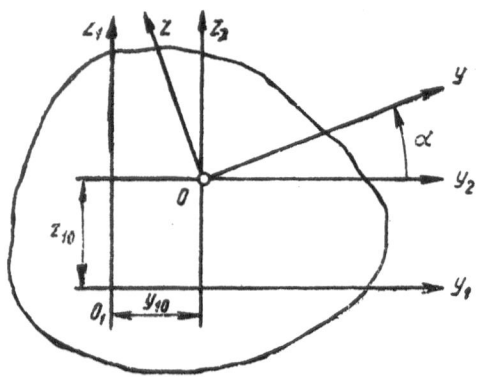

Fig. 17. Finding the coordinates of the reduced center of gravity.

The components of the shear deformation of the loading $-\varepsilon_s = \varepsilon_x^s - \varepsilon + \alpha t$ and γ_{xy}^s are, from Hooke's law, proportional to the corresponding stresses, i.e.,

$$\frac{\varepsilon_x^s - \varepsilon_s}{\sigma_x - \sigma} = \frac{\varepsilon_y^s - \varepsilon_s}{\sigma_y - \sigma} = \frac{\varepsilon_z^s - \varepsilon_s}{\sigma_z - \sigma} = \frac{\gamma_{xy}^s}{2\sigma_{xy}} = \frac{\gamma_{yz}^s}{2\sigma_{yz}} = \frac{\gamma_{zx}^s}{2\sigma_{zx}} = \frac{1}{2G}. \tag{4.50}$$

From the conditions (4.47) and (4.50), we have

$$\sigma_x - \sigma = 2G \left(\varepsilon_x - \varepsilon_x^p - \varepsilon \right),$$
$$\sigma_{xy} = G \left(\gamma_{xy} - \gamma_{xy}^p \right),$$

and, expressing σ in terms of θ from Eq. (4.49), we obtain

$$\left. \begin{array}{c} \sigma_x = 2G \left(\varepsilon_x + \dfrac{\mu\theta}{1 - 2\mu} - \varepsilon_x^p - \dfrac{1 + \mu}{1 - 2\mu} \alpha t \right), \\ \cdots\cdots\cdots\cdots\cdots\cdots\cdots\cdots\cdots \\ \sigma_{xy} = G \left(\gamma_{xy} - \gamma_{xy}^p \right), \\ \cdots\cdots\cdots\cdots\cdots\cdots \end{array} \right\} \tag{4.51}$$

and, inversely,

$$\left. \begin{array}{c} \varepsilon_x = \dfrac{1}{2G} \left(\sigma_x - \dfrac{3\mu}{1 + \mu} \sigma \right) + \varepsilon_x^p + \alpha t, \\ \cdots\cdots\cdots\cdots\cdots\cdots\cdots\cdots\cdots \\ \gamma_{xy} = \dfrac{\sigma_{xy}}{G} + \gamma_{xy}^p, \\ \cdots\cdots\cdots\cdots \end{array} \right\} \tag{4.52}$$

For $\varepsilon_x^P = 0$ and $\gamma_{xy}^P = 0$, Eqs. (4.51)-(4.52) reduce to the ordinary equations of elasticity theory.

If the components of the creep deformation are known at a given instant of time, Eqs. (4.51)-(4.52), together with the equilibrium equations (3.40), the boundary conditions (3.41), and the geometric relations

$$\left. \begin{array}{ll} \varepsilon_x = \dfrac{\partial u}{\partial x} & \gamma_{xy} = \dfrac{\partial u}{\partial y} + \dfrac{\partial v}{\partial x}, \\[2mm] \varepsilon_y = \dfrac{\partial v}{\partial y} & \gamma_{yz} = \dfrac{\partial v}{\partial z} + \dfrac{\partial w}{\partial y}, \\[2mm] \varepsilon_z = \dfrac{\partial w}{\partial z} & \gamma_{zx} = \dfrac{\partial w}{\partial x} + \dfrac{\partial u}{\partial z}, \end{array} \right\} \tag{4.53}$$

where u, v, and w are the displacements in the directions of the X, Y, and Z axes, make it possible to use the ordinary methods of elasticity theory to find the stresses.

For a number of problems, the solution may be written in the form

$$\sigma_x = \sigma_x^0 + L_x \left(\varepsilon_x^p, \varepsilon_y^p, \ldots \right), \tag{4.54}$$

where σ_x^0 is the value of σ_x for $\varepsilon_x^P = \varepsilon_y^P = \ldots = 0$, and L is some linear integral operator. This in particular is the form in which the problems of tension and bending of a rod were solved above [16].

It may be seen from Eqs. (4.51)-(4.52) that the effect of creep deformations on the stressed state is similar to the effect of certain temperature deformations (in an anisotropic material). Therefore, the problem of finding σ from ε^P may usually be solved in the same way as in the corresponding thermoelastic problem.

In order to find the values of ε_x^p at a given instant of time, the problem must be solved in successive stages, in a way similar to what was done in the uniaxial problems discussed above.

Since, from Eq. (4.4),

$$\frac{\partial p}{\partial \tau} = v_{pi}, \tag{4.55}$$

Eqs. (4.1) are integrated in the same form as in the tension problem, i.e., Eqs. (4.18)-(4.21) continue to hold for the total stressed state if $|\sigma|$ is replaced by σ_i.

The components of the instantaneous plastic deformation are found from Eq. (4.5)

$$\left.\begin{aligned}
\varepsilon_{x,\,n+1}^p &= \varepsilon_{x,\,n}^p + \frac{3}{2} \int_{\tau_n}^{\tau_{n+1}} \frac{v_{pi}}{\sigma_i} (\sigma_x - \sigma)\, d\tau, \\
&\qquad \cdots\cdots\cdots\cdots \\
\gamma_{xy,\,n+1}^p &= \gamma_{xy,\,n}^p + 3 \int_{\tau_n}^{\tau_{n+1}} \frac{v_{pi}}{\sigma_i} \sigma_{xy}\, d\tau \\
&\qquad \cdots\cdots\cdots\cdots
\end{aligned}\right\} \tag{4.56}$$

where

$$\sigma = \frac{1}{3} (\sigma_x + \sigma_y + \sigma_z).$$

If it is assumed that the nature of the stressed state does not change during the time $\Delta \tau_n$ (the Mohr circles remain similar), i.e.,

$$\left. \frac{\sigma_x - \sigma}{\sigma_i} \right|_{n+1} \approx \left. \frac{\sigma_x - \sigma}{\sigma_i} \right|_n \; ; \quad \left. \frac{\sigma_{xy}}{\sigma_i} \right|_{n+1} \approx \left. \frac{\sigma_{xy}}{\sigma_i} \right|_n , \tag{4.57}$$

then, from Eq. (4.56), using (4.55), we obtain

$$\left.\begin{aligned}
\varepsilon_{x,\,n+1}^p &= \varepsilon_{x,\,n}^p + \frac{3(\sigma_{xn} - \sigma_n)}{2\sigma_{in}} (p_{n+1} - p_n), \\
&\qquad \cdots\cdots\cdots\cdots\cdots \\
\gamma_{xy,\,n+1}^p &= \gamma_{xy,\,n}^p + \frac{3\sigma_{xy,\,n}}{\sigma_{in}} (p_{n+1} - p_n).
\end{aligned}\right\} \tag{4.58}$$

Since $\sigma = \sigma_0/3$ and $\sigma_i = |\sigma_0|$ for a uniaxially stressed state, Eqs. (4.58) reduce in this case to (4.22). Thus, if at the instant of time τ_n we know the stresses $\sigma_{xn}, \ldots, \sigma_{xy,n}, \ldots$, the instantaneous deformations $\varepsilon_{xn}, \ldots, \gamma_{xy,n}^p, \ldots$, and the accumulated plastic deformation p_n, then, fixing the time interval $\Delta \tau_n$, we find for each point in the body the deformations p_{n+1}, and then $\varepsilon_{x,\,n+1}^p, \ldots, \gamma_{xy,\,n+1}^p$. Further, from equations of the type (4.54), we find $\sigma_{x,\,n+1}$ and $\sigma_{xy,\,n+1}$, after which we can pass to the next stage of the calculation, as long as $\Delta \tau_n$ was chosen such that the condition of type (4.19) with $|\sigma|$ replaced by σ_i and L replaced by L_i, is satisfied.

Hollow Nonuniformly Heated Cylinder

A discussion was given in Section 3 of the problem of stabilized creep in a nonuniformly heated cylinder under internal and external pressures and an axial force. In order to get a solution in the unstabilized creep range according to the method given in the previous section, we find the expressions for the operators L, which establish a relation between the stresses and the fixed creep deformations.

If w is the displacement in the radial direction, it follows from geometric considerations that

$$\varepsilon_r = \frac{dw}{dr} , \quad \varepsilon_\theta = \frac{w}{r} . \tag{4.59}$$

154

In a very long cylinder, the longitudinal elongation is ε_x = const. Then

$$\theta = \varepsilon_r + \varepsilon_\theta + \varepsilon_x = \frac{dw}{dr} + \frac{w}{r} + \varepsilon_x.$$

and Eqs. (4.51) take the form

$$\begin{aligned}
\sigma_r &= 2G\left[\frac{dw}{dr} + \frac{\mu}{1-2\mu}\left(\frac{dw}{dr} + \frac{w}{r} + \varepsilon_x\right) - \varepsilon_r^p - \frac{1+\mu}{1-2\mu}\,\alpha t\right], \\
\sigma_\theta &= 2G\left[\frac{w}{r} + \frac{\mu}{1-2\mu}\left(\frac{dw}{dr} + \frac{w}{r} + \varepsilon_x\right) - \varepsilon_\theta^p - \frac{1+\mu}{1-2\mu}\,\alpha t\right], \\
\sigma_x &= 2G\left[\varepsilon_x + \frac{\mu}{1-2\mu}\left(\frac{dw}{dr} + \frac{w}{r} + \varepsilon_x\right) - \varepsilon_x^p - \frac{1+\mu}{1-2\mu}\,\alpha t\right].
\end{aligned} \tag{4.60}$$

From symmetry, all the tangential components of the stresses and deformations are equal to zero.

Substituting Eq. (4.60) in the equation for radial equilibrium of an element of the cylinder (3.52), we obtain

$$2\frac{d}{dr}\left\{G\left[\frac{dw}{dr} + \frac{\mu}{1-2\mu}\left(\frac{dw}{dr} + \frac{w}{r} + \varepsilon_x\right) - \varepsilon_r^p - \frac{1+\mu}{1-2\mu}\,\alpha t\right]\right\} - \frac{2G}{r}\left(\frac{w}{r} - \frac{dw}{dr} - \varepsilon_\theta^p + \varepsilon_r^p\right) = 0. \tag{4.61}$$

In the general case, the modulus G changes with temperature. However, usually, for a temperature change of the order of 100-200°C, the change in G is not greater than 10-15% (if the temperature is not too high). Since using the relation $G = f(T)$ complicates the calculation substantially, we take for simplicity $G = G_m$ = const. Then it follows from Eq. (4.61) that

$$\frac{1-\mu}{1-2\mu}\left(\frac{d^2 w}{dr^2} + \frac{1}{r}\frac{dw}{dr} - \frac{w}{r^2}\right) = \frac{d\varepsilon_r^p}{dr} - \frac{\varepsilon_\theta^p - \varepsilon_r^p}{r} + \frac{1+\mu}{1-2\mu}\frac{d}{dr}(\alpha t),$$

or

$$\frac{d}{dr}\left[\frac{1}{r}\frac{d}{dr}(rw)\right] = \frac{1-2\mu}{1-\mu}\left[\frac{d\varepsilon_r^p}{dr} - \frac{\varepsilon_\theta^p - \varepsilon_r^p}{r}\right] + \frac{1+\mu}{1-\mu}\frac{d}{dr}(\alpha t). \tag{4.62}$$

Integrating Eq. (4.62) twice, we obtain

$$w = \frac{b_1}{r} + b_2 r + \frac{1-2\mu}{1-\mu}\frac{1}{r}\left[\int_a^r r_1 \varepsilon_r^p\, dr_1 - \int_a^r r_1 \int_a^{r_1} \frac{\varepsilon_\theta^p - \varepsilon_r^p}{r^2}\, dr_2 dr_1\right] + \frac{1+\mu}{1-\mu}\frac{1}{r}\int_a^r \alpha t r_1 dr_1, \tag{4.63}$$

where b_1 and b_2 are integration constants.

We transform the expression in the square brackets, taking the second integral by parts:

$$\frac{1}{r}\left[\int_a^r r_1\varepsilon_r^p dr_1 - \int_a^r r_1 \int_a^r \frac{\varepsilon_\theta^p - \varepsilon_r^p}{r_2}\, dr_2 dr_1\right] = \frac{r}{2}\left[\frac{1}{r^2}\int_a^r r_1(\varepsilon_\theta^p + \varepsilon_r^p)dr_1 - \int_a^r \frac{\varepsilon_\theta^p - \varepsilon_r^p}{r_1}\, dr_1\right]. \tag{4.64}$$

We introduce the notation

$$\begin{aligned}
A(r) &= \frac{2G}{r^2}\int_a^r r_1(\varepsilon_\theta^p + \varepsilon_r^p)\, dr_1, \\
B(r) &= 2G\int_a^r \frac{1}{r_1}(\varepsilon_\theta^p - \varepsilon_r^p)\, dr_1.
\end{aligned} \tag{4.65}$$

It is easily seen that the integral expressions A(r) and B(r) have the dimensions of stress and are equal to zero at r = a. Making use of Eqs. (4.64) and (4.65), the expression for the displacement w takes the form

$$w = \frac{b_1}{r} + b_1 r + \frac{1-2\mu}{1-\mu} \frac{r}{4G} [A(r) - B(r)] + \frac{1+\mu}{1-\mu} \frac{1}{r} \int_a^r \alpha t r_1 \, dr_1. \tag{4.66}$$

In order to find the values of b_1 and b_2 from the boundary conditions $\sigma_r(a) = -p_a$, and $\sigma_r(b) = -p_b$, we express σ_r from Eqs. (4.60) and (4.66) in the form:

$$\sigma_r = 2G \left(\frac{1-\mu}{1-2\mu} \frac{dw}{dr} + \frac{\mu}{1-2\mu} \frac{w}{r} + \frac{\mu}{1-2\mu} \varepsilon_x - \varepsilon_r^p - \frac{1+\mu}{1-2\mu} \alpha t \right)$$

$$= \frac{2G}{1-2\mu} \left\{ (1-\mu) b_2 - (1-\mu) \frac{b_1}{r^2} \right.$$

$$+ \frac{1-2\mu}{4G} \left[A(r) - B(r) + r \frac{dA}{dr} - r \frac{dB}{dr} \right] - (1+\mu) \frac{1}{r^2} \int_a^r \alpha t r_1 \, dr_1$$

$$+ (1+\mu) \alpha t + \frac{\mu b_1}{r^2} + \mu b_2 + \frac{1-2\mu}{1-\mu} \frac{\mu}{4G} [A(r) - B(r)]$$

$$+ \frac{1+\mu}{1-\mu} \frac{\mu}{r^2} \int_a^r \alpha t r_1 \, dr_1 + \mu \varepsilon_x - (1-2\mu) \varepsilon_r^p - (1+\mu) \alpha t \right\}$$

$$= \frac{2G}{1-2\mu} \left[b_2 - (1-2\mu) \frac{b_1}{r^2} + \mu \varepsilon_x \right] + \frac{1}{2(1-\mu)} [A(r) - B(r)]$$

$$+ \frac{r}{2} \left(\frac{dA}{dr} - \frac{dB}{dr} \right) - 2G \varepsilon_r^p - \frac{2G(1+\mu)}{1-\mu} \frac{1}{r^2} \int_a^r \alpha t r_1 \, dr_1.$$

Remembering that from Eqs. (4.65)

$$\begin{cases} \dfrac{dA}{dr} = -\dfrac{2}{r} A(r) + \dfrac{2G}{r} (\varepsilon_\theta^p + \varepsilon_r^p), \\[2mm] \dfrac{dB}{dr} = \dfrac{2G}{r} (\varepsilon_\theta^p - \varepsilon_r^p), \end{cases} \tag{4.67}$$

we obtain:

$$\sigma_r = \frac{2G}{1-2\mu} \left[b_2 - (1-2\mu) \frac{b_1}{r^2} + \mu \varepsilon_x \right] - \frac{1}{2(1-\mu)} [(1-2\mu) A(r) + B(r)] - \frac{2G(1+\mu)}{1-\mu} \frac{1}{r^2} \int_a^r \alpha t r_1 \, dr_1. \tag{4.68}$$

At the inner radius with r = a,

$$-p_a = \frac{2G}{1-2\mu} \left[b_2 - (1-2\mu) \frac{b_1}{a^2} + \mu \varepsilon_x \right]. \tag{4.69}$$

At the outer radius with r = b,

$$-p_b = \frac{2G}{1-2\mu} \left[b_2 - (1-2\mu) \frac{b_1}{b^2} + \mu \varepsilon_x \right] - \frac{1}{2(1-\mu)} [(1-2\mu) A(b) + B(b)] - \frac{2G(1+\mu)}{1-\mu} \frac{1}{b^2} \int_a^b \alpha t r \, dr.$$

Solving Eq. (4.69), we find the values of the constants b_1 and b_2 expressed in terms of the still unknown axial deformation ε_x. This latter quantity is found from the axial equilibrium equation (3.56), where the expression for σ_x must be put in as given by Eqs. (4.60) and (4.66).

After some simple but rather unwieldy manipulations, which we do not give here, we come to the following expressions for the stresses:

$$\sigma_r = \sigma_r^N + \sigma_r^t + \sigma_r^p, \tag{4.70}$$

where $\sigma_r^N, \dots$ are the stresses from external forces found by the familiar Lamé equations:

$$
\left.
\begin{aligned}
\sigma_r^N &= \frac{a^2}{1-\alpha^2}\, P_a \left(1 - \frac{b^2}{r^2} \right) - \frac{1}{1-\alpha^2}\, P_b \left(1 - \frac{a^2}{r^2} \right), \\
\sigma_\theta^N &= \frac{a^2}{1-\alpha^2}\, P_a \left(1 + \frac{b^2}{r^2} \right) - \frac{1}{1-\alpha^2}\, P_b \left(1 + \frac{a^2}{r^2} \right), \\
\sigma_x^N &= \frac{P_0 + \pi b^2 (P_a a^2 - P_b)}{\pi b^2 (1-\alpha^2)},
\end{aligned}
\right\} \tag{4.71}
$$

where $\alpha = a/b$, σ_r^t are the thermal stresses, also found from the widely known equations

$$
\left.
\begin{aligned}
\sigma_r^t &= \frac{E}{(1-\mu)\, b^2} \left[\frac{1}{1-\alpha^2} \left(1 - \frac{a^2}{r^2} \right) \int_a^b \alpha t r\, dr - \frac{b^2}{r^2} \int_a^r \alpha t r_1\, dr_1 \right], \\[2mm]
\sigma_\theta^t &= \frac{E}{(1-\mu)\, b^2} \left[\frac{1}{1-\alpha^2} \left(1 + \frac{a^2}{r^2} \right) \int_a^b \alpha t r\, dr + \frac{b^2}{r^2} \int_a^r \alpha t r_1\, dr_1 - b^2 \alpha t \right], \\[2mm]
\sigma_x^t &= \frac{E}{(1-\mu)\, b^2} \left[\frac{2}{1-\alpha^2} \int_a^b \alpha t r\, dr - b^2 \alpha t \right],
\end{aligned}
\right\} \tag{4.72}
$$

and, finally, $\sigma_r^p \dots$ are the stresses depending solely on the fixed values of the components of the creep deformation. Comparing Eqs. (4.70) and (4.54) shows that

$$
\left.
\begin{aligned}
\sigma_r^0 &= \sigma_r^N + \sigma_r^t, \\
&\cdots\cdots\cdots \\
L_r &= \sigma_r^p. \\
&\cdots\cdots\cdots
\end{aligned}
\right\} \tag{4.73}
$$

etc.

The formulas used to calculate the operators L_r, L_θ, and L_x are of the following form:

$$
L_r = \frac{(1-2\mu)\, A(b)}{2(1-\mu)(1-\alpha^2)} \left(1 - \frac{a^2}{r^2} \right) \Bigg| + \frac{B(b)}{2(1-\mu)(1-\alpha^2)} \left(1 - \frac{a^2}{r^2} \right) - \frac{1-2\mu}{2(1-\mu)} A(r) \Bigg| - \frac{1}{2(1-\mu)} B(r),
$$

$$
L_\theta = \frac{(1-2\mu)\, A(b)}{2(1-\mu)(1-\alpha^2)} \left(1 + \frac{a^2}{r^2} \right) \Bigg| + \frac{B(b)}{2(1-\mu)(1-\alpha^2)} \left(1 + \frac{a^2}{r^2} \right) + \frac{1-2\mu}{2(1-\mu)} A(r) \tag{4.74}
$$
$$
- \frac{1}{2(1-\mu)} B(r) + \frac{\mu}{1-\mu} 2G\varepsilon_r^p(r) - 2G\varepsilon_\theta^p(r),
$$

$$
L_x = \frac{2D}{1-\alpha^2} - \frac{\mu A(b)}{(1-\mu)(1-\alpha^2)} + \frac{\mu B(b)}{(1-\mu)(1-\alpha^2)} \Bigg| - \frac{\mu B(r)}{1-\mu} + \frac{\mu}{1-\mu} 2G\varepsilon_r^p(r) - 2G\varepsilon_x(r),
$$

where $D = \dfrac{2G}{b^2} \displaystyle\int\limits_{a}^{b} r\varepsilon_x^p \, dr$, A(r), and B(r) are found from Eqs. (4.65), and A(b) and B(b) are equal to the values

of A(r) and B(r) at r = b.

Equations (4.74), together with the creep equations, permit making a complete unstabilized creep calculation for a nonuniformly heated cylinder.

LITERATURE CITED

1. Lebedev, N. N., Thermal Stresses in the Theory of Elasticity, Moscow-Leningrad, ONTI, 1937.
2. Maizel', V. M., The Temperature Problem of the Theory of Elasticity, Izd. AN UkrSSR, 1951.
3. Melan, É., and Parkus, G., Thermoelastic Stresses Produced by Steady-State Temperature Fields (translation from German) Moscow, Fizmatgiz, 1958.
4. Kachanov, L. M.,"Elastoplastic equilibrium of nonuniformly heated thin-walled cylinders under internal pressure," Zhur. Tekhn. Fiz. 10 (1940).
5. Gatewood, B. E., Thermal Stresses (translation from English) Moscow, IL, 1959.
6. Rabotnov, Yu. N., "Possibilities of describing unstabilized creep with application to the study of creep," Izv. Akad. Nauk SSSR, Otd. Tekhn. Nauk, No. 5 (1957).
7. Malinin, N. N., Fundamentals of Creep Calculations, Moscow, Mashgiz, 1948.
8. Bailey, R. W., "The utilization of creep test data in engineering design," Proc. Inst. Mech. Engs., Vol. 131 (1935).
9. Kachanov, L. M., Some Problems in the Theory of Creep, Moscow, Gostekhizdat, 1949.
10. Ponomarev, S. D., et al., Fundamentals of Modern Methods of Calculating for Strength in Machine Construction, Col. 2, Moscow, Mashgiz, 1952.
11. Rabotnov, Yu. N., "Designing machine parts for creep," Izv. Akad. Nauk. Otd. Tekhn. Nauk, No. 6 (1948).
12. Shorr, B. V., "Effect of nonuniform heating under creep conditions on the change in the stressed state," Dokl. Akad. Nauk SSSR 123, No. 5 (1958).
13. Shesterikov, S. A., "One conditions for creep laws," Izv. Akad. Nauk SSSR, Otd. Techn. Nauk Mekh. i Mashinostr., No. 1 (1959).
14. Mikheev, M. A., Fundamentals of Heat Transfer, Moscow-Leningrad, Gosénergoizdat, 1949.
15. Timoshenko, S. P., Strength of Materials, Part 2, Moscow, Gostekhizdat, 1946.
16. Shorr, B. F., "Unstabilized creep calculations on nonuniformly heated rods of arbitrary cross section" Izv. Akad. Nauk Otd. Tekhn. Nauk, Mekh. i Mashinostr., No. 1 (1959).

THERMAL STABILITY OF PLATES AND SHELLS

L. A. Shapovalov

The demands of many modern branches of engineering are such that questions dealing with loss of stability of structural elements at elevated temperatures have recently begun to attract the attention of large groups of engineers and investigators.

This review gives a very compact presentation of some of the results of Russian and foreign work that has been done on thermal buckling of plates and shells, so as to acquaint engineers engaged in designing structural elements of this type with present-day methods of designing for stability under heating.

Introduction

One of the basic methods of calculating the critical temperatures corresponding with the instant at which an elastic solid buckles in a given temperature field is the one based on the extremum properties of the potential energy built up by the system as a result of the thermal stresses produced by heating.

In neutral equilibrium, i.e., in the critical state,

$$\delta\Pi = 0; \qquad \delta^2\Pi = 0; \tag{1}$$

$$\Pi = \int\int\int \frac{1}{2E}\left[(\sigma_{xx}^2 + \sigma_{yy}^2 + \sigma_{zz}^2) - 2\nu(\sigma_{xx}\sigma_{yy} + \sigma_{yy}\sigma_{zz} + \sigma_{zz}\sigma_{xx}) + 2(1+\nu)(\sigma_{xy}^2 + \sigma_{yz}^2 + \sigma_{zx}^2)\right]dx\,dy\,dz, \tag{2}$$

where Π is the potential energy of the system, and σ_{xx}, σ_{yy}, ... , σ_{zx} are the thermal stresses.

In this sort of purely mechanical treatment of thermal buckling, finding the critical temperatures reduces to calculating the temperature stresses, and then finding the stability of the forms of equilibrium that are possible under given stresses from the conditions (1) or by using other methods.

Some authors, and in particular Hoff [1], do not regard this method of investigating the buckling as rigorous when applied to the temperature problem, since the thermal stresses are not produced by external loads, but by thermal expansion, and depend on the deformations in the solid when it has lost stability.

In our opinion, the most rigorous solution of the thermal buckling problem can only be found by thermodynamic methods, where the heated elastic solid is regarded as a thermodynamic system. The critical values of the system parameters may be found from the conditions for thermodynamic equilibrium in the system, i.e., from the extremum principles for the thermodynamic potentials in question.

Thermal Buckling of Flat and Curved Plates

Following Hoff's paper [1], consider a flat elastic isotropic plate of infinite length, freely supported on the long sides, and compressed by thermal stresses in the longitudinal direction (Fig. 1).

The differential equation for the bending of the plate leads to a linear partial differential equation with variable coefficients, which may be written in the form

$$\nabla^4 w = -\frac{h}{D}\sigma w_{,yy}, \tag{3}$$

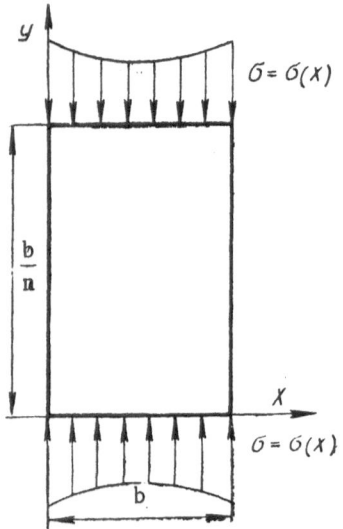

Fig. 1. Loading scheme of a clamped plate (conventional notation).

where h is the thickness of the plate, w is the deflection (subscripts coming after the comma indicate differentiation), and D is the cylindrical rigidity of the plate,

$$D = \frac{Eh^3}{12(1-\nu^2)};$$

and ν is Poisson's coefficient.

The operator $\nabla^4 w$ is of the form

$$\nabla^4 w = \frac{\partial^4 w}{\partial x^4} + 2\frac{\partial^4 w}{\partial x^2 \partial y^2} + \frac{\partial^4 w}{\partial y^4}. \tag{4}$$

The boundary conditions for the problem are:

$$x = 0, b \qquad w = \frac{\partial^2 w}{\partial x^2} = 0$$

$$y = 0, \frac{b}{n} \qquad w = \frac{\partial^2 w}{\partial y^2} = 0,$$

where b/n is the wavelength in the longitudinal direction. The compressive stress σ is variable over the width of the plate.

Without going through the problem of calculating the thermal stresses, Hoff [1] assumes the above stresses, corresponding with some temperature field symmetric and constant over the thickness, to be given by the trigonometric series:

$$\sigma = \frac{\sigma_0}{c} \sum_{m=0}^{\infty} p_m \cos\frac{m\pi x}{b}, \tag{5}$$

where

$$\sigma_0 = \frac{\pi^2 E}{12(1-\nu^2)}\left(\frac{h}{b}\right)^2.$$

The curved state of the plate after buckling is given in the form

$$w = \sin\frac{n\pi y}{b} \sum_{k=1}^{\infty} a_k \sin\frac{k\pi x}{b}. \tag{6}$$

By substituting the expression for σ and w in the original differential equation for the bendings (3), the buckling condition may be found by equating to zero the infinite determinate

$$\begin{vmatrix} 2(p_0-ck_1)-p_2 & p_1-p_3 & p_2-p_4 \dots \\ p_1-p_3 & 2(p_0-ck_2)-p_4 & p_1-p_5 \dots \\ p_2-p_4 & p_1-p_5 & 2(p_0-ck_3)-p_6 \dots \\ p_3-p_5 & p_2-p_6 & p_1-p_7 \dots \\ \dots & \dots & \dots \end{vmatrix} = 0, \tag{7}$$

$$k_i = \left(\frac{h^2+i^2}{h}\right).$$

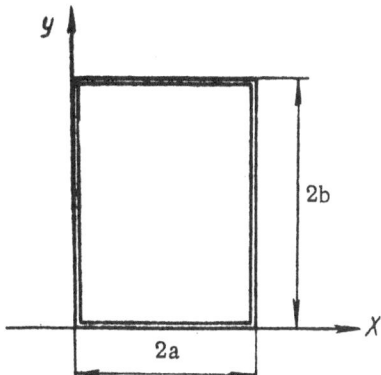

Fig. 2. Freely supported plate (conventional notation).

Equation (7) gives the critical value of the parameter c. In uniform compression, i.e., where the thermal stresses are constant over the width of the plate,

$$p_0 = 1, \qquad p_m = 0 \quad \text{for} \quad m \neq 0,$$

the minimum critical buckling stresses are

$$\sigma_{0\,\mathrm{cr}} = 4\sigma_0.$$

For a nonuniform distribution of thermal stresses, Eq. (7) is solved numerically.

The paper by Klosner and Forray [2] gives a discussion of thermal buckling of plates of finite dimensions, supported on elastic elements. It is assumed that the supporting elements have no torsional rigidity. The plate and the supports are assumed to be made of different materials, and hence thermal stresses may occur even with uniform heating. The physical constants of the plate and the elastic supports are calculated for the mean temperatures of the parts in question, and are assumed constant.

The temperature field of the plate is given by the authors in the form

$$T = \overline{T} + T_0 + \sum_{s=0}^{\infty} \sum_{t=0}^{\infty} T_{st} \cos\left(\frac{s\pi x}{a}\right) \cos\left(\frac{t\pi y}{b}\right), \tag{8}$$

where $\overline{T}$ is the difference between the mean temperature of the supports and the ambient temperature; T_0 is the difference between the temperature of the edges of the plates and the mean temperature of the supports, and 2a and 2b are the dimensions of the plate in plan (Fig. 2).

The thermal stresses produced by the given symmetric temperature field (8) are calculated from the equation

$$\nabla^4 \Psi = -E\alpha\nabla^2 T, \tag{9}$$

where Ψ is the stress function,

$$\frac{\partial^2 \Psi}{\partial y^2} = \sigma_x, \qquad \frac{\partial^2 \Psi}{\partial x^2} = \sigma_y, \qquad \frac{\partial^2 \Psi}{\partial x \partial y} = -\tau_{xy},$$

$$\nabla^2 T = \frac{\partial^2 T}{\partial x^2} + \frac{\partial^2 T}{\partial y^2};$$

and α is the thermal coefficient of linear expansion.

The longitudinal thermal loads N_x, N_y, and N_{xy} are found from the resulting values of the stresses, and the potential energy is set up for the plate in the deformed state

$$\Pi = \frac{D}{2} \int_0^{2a} \int_0^{2b} \left[\left(\frac{\partial^2 w}{\partial x^2} + \frac{\partial^2 w}{\partial y^2} \right)^2 - 2(1-\nu) \left\{ \frac{\partial^2 w}{\partial x^2} \frac{\partial^2 w}{\partial y^2} \right. \right.$$

$$\left. \left. - \left(\frac{\partial^2 w}{\partial x \partial y} \right)^2 \right\} \right] dx\,dy + \frac{1}{2} \int_0^{2a} \int_0^{2b} \left[N_x \left(\frac{\partial w}{\partial x} \right)^2 \right.$$

$$\left. + 2N_{xy} \left(\frac{\partial w}{\partial x} \frac{\partial w}{\partial y} \right) + N_y \left(\frac{\partial w}{\partial y} \right)^2 dx\,dy. \right. \tag{10}$$

161

The function which approximates the deflection and satisfies the freely supported boundary conditions is written as the double trigonometric series:

$$w = \sum_{m=1,3}^{\infty} \sum_{n=1,3}^{\infty} a_{mn} \sin \frac{2\pi x}{\lambda_m} \sin \frac{2\pi_y}{\mu_n}. \tag{11}$$

In Eq. (11), the authors limit themselves to the first four terms of the series. Substituting the function (11) into Eq. (10), and making use of condition (1) gives the following characteristic equation for finding the critical temperature:

$$\begin{vmatrix} K_{11} & K_{13} & K_{31} & K_{33} \\ L_{11} & L_{13} & L_{31} & L_{33} \\ M_{11} & M_{13} & M_{31} & M_{33} \\ N_{11} & N_{13} & N_{31} & N_{33} \end{vmatrix} = 0, \tag{12}$$

where K, L, N, and M are quantities depending on the parameters of the plate, the coefficients λ_m, μ_n, and the temperature field [2].

The example is discussed of a plate held on rigid supports which allow free rotation of the edges, the plate being heated by the temperature field

$$T = T_0 + T_1 \{1 - [(x - a)/a]^2\} \{1 - [(y - b)/b]^2\}, \tag{13}$$

where $\dot{T}_1$ is the difference between the temperature at the center of the plate and at the edges.

Here, the following value is found for the critical temperature T_1 at the center of the plate:

$$T_{1_{\text{cr}}} = \frac{k_t}{1 - \nu^2} \frac{1}{a} \left(\frac{h}{b}\right)^2. \tag{14}$$

The coefficient k_t depends on the ratio of the sides of the plate, and the temperature ratio T_0/T_1 (Fig. 3).

Figure 3 also gives values of the coefficient k_t for a uniform temperature field and for the mean temperatures of the plate.

It is not difficult to see that the nonuniformity of the heating may be neglected for temperature drops which have the ratio $T_0/T_1 \geq 2$. At the same time, the nonuniformity exerts a substantial effect for $T_0/T_1 < 2$. The maximum effect of nonuniform heating occurs for long plates. In this case, the critical temperatures are about 20% less than for uniform heating.

In nonuniform heating, where the expansion of some zones prevents the expansion of others, thermal stresses occur in the solid. It is obvious that temperature stresses also occur in uniform heating if the solid is clamped in a definite way.

Consider a circularly cylindrical panel (Fig. 4) of constant thickness h, length 2a, width 2b, and radius R, in the linear temperature field

$$T_0 = t_0(x, y, t) + \frac{z}{h} t_1(x, y, t), \tag{15}$$

where t is a temperature parameter.

M. S. Ganeeva [3] reduces the problem of the stability of a curvilinear panel of this sort in the temperature field (15) to the following system of differential equations:

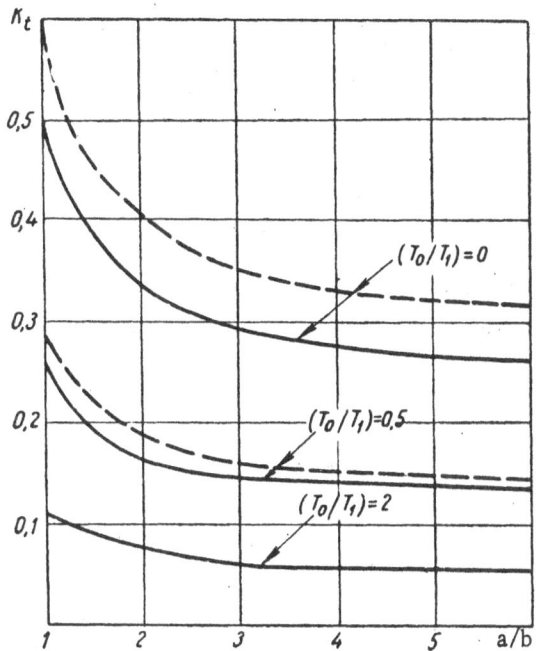

Fig. 3. Critical temperature parameter as a function of the ratio of the sides of the plate: solid curves — nonuniform heating; dotted curves — mean temperatures.

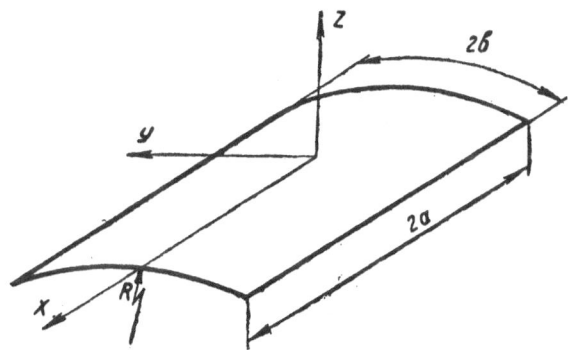

Fig. 4. Curvilinear plate (conventional notation).

$$\nabla\nabla\Phi + l\,(1-\nu)\,\nabla t_0 = Eh\left[\left(\frac{\partial^2 w}{\partial x\partial y}\right)^2 - \frac{\partial^2 w}{\partial x^2}\frac{\partial^2 w}{\partial y^2} + \frac{1}{R}\frac{\partial^2 w}{\partial x}\right];$$

$$\mathrm{D}\,\nabla\nabla w + m\nabla t_1 = \frac{\partial^2\Phi}{\partial y^2}\frac{\partial^2 w}{\partial x} - \left(\frac{1}{R} - \frac{\partial^2 w}{\partial y^2}\right)\frac{\partial^2\Phi}{\partial x^2} - 2\frac{\partial^2 w}{\partial x\partial y}\frac{\partial^2\Phi}{\partial x\partial y} \qquad (16)$$

where Φ is the stress function,

$$m = \frac{\alpha Eh^2}{12\,(1-\nu)};$$

∇ is Laplace's operator, and $\nabla\nabla$ is the biharmonic operator.

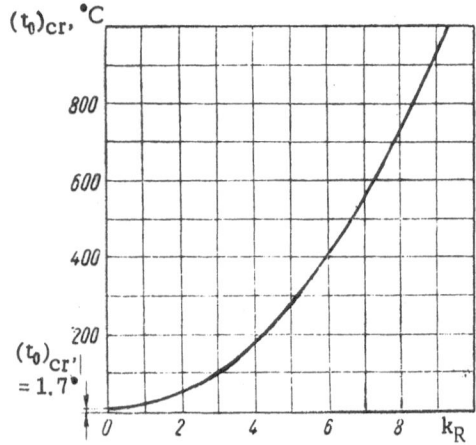

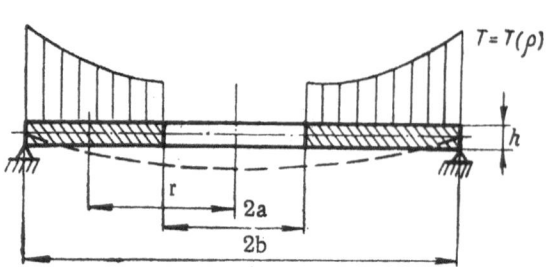

Fig. 5. **Critical temperature of a curvi-linear plate as a function of the curvature parameter.**

Fig. 6. **Circular plate (conventional notation).**

For a steady-state temperature field, $\nabla t_1 = \nabla t_0 = 0$, and Eq. (16) becomes the same as the corresponding equations for the nontemperature problem. In this case, the effect of the temperature field will be given by the boundary conditions.

Consider a special case – the uniform temperature field of a clamped panel, $T = t_0 = $ const. The boundary conditions are written in the form

$$\text{for} \quad x = \pm a \quad u = v = w = \frac{\partial w}{\partial x} = 0,$$

$$\text{for} \quad y = \pm b \quad u = v = w = \frac{\partial w}{\partial y} = 0,$$

where u and v are the tangential displacements.

Expressing the tangential displacements in terms of the stress function and the displacement function, we obtain the boundary conditions in the following form:

$$\left.\begin{array}{l}
\text{for} \quad x = \pm a \;\; w = 0, \; \dfrac{\partial w}{\partial x} = 0, \dfrac{\partial^2 \Phi}{\partial x^2} - \nu \dfrac{\partial^2 \Phi}{\partial y^2} + Eh\alpha t_0 = 0, \\[2mm]
\displaystyle\int_0^{\pm a} \left\{ \dfrac{\partial^2 \Phi}{\partial y^2} - \nu \dfrac{\partial^2 \Phi}{\partial x^2} + Eh \left[\alpha t_0 - \dfrac{1}{2} \left(\dfrac{\partial w}{\partial x} \right)^2 \right] \right\} dx = 0, \\[4mm]
\text{for} \quad y = \pm b \;\; w = 0, \dfrac{\partial w}{\partial y} = 0, \quad \dfrac{\partial^2 \Phi}{\partial y} - \nu \dfrac{\partial^2 \Phi}{\partial x^2} + Eh\alpha t_0 = 0, \\[2mm]
\displaystyle\int_0^{\pm b} \left\{ \dfrac{\partial^2 \Phi}{\partial x^2} - \nu \dfrac{\partial^2 \Phi}{\partial y^2} + Eh \left[\alpha t_0 - \dfrac{w}{R} - \dfrac{1}{2} \left(\dfrac{\partial w}{\partial y} \right)^2 \right] \right\} dy = 0.
\end{array}\right\} \quad (17)$$

M. S. Ganeeva uses the Bubnov-Galerkin method for solving the problem, assuming in the first approximation

$$w = w_0 \left(1 + \cos \frac{\pi x}{a} \right) \left(1 + \cos \frac{\pi y}{b} \right), \tag{18}$$

and defining the stress function Φ in the form

$$\Phi = C_1 E_1(x) \cos \frac{\pi y}{b} + D_1 E_1(y) \cos \frac{\pi x}{a} + \gamma_1 \operatorname{ch} \frac{\pi x}{b} \cos \frac{\pi y}{b}$$

$$+ \delta_1 \operatorname{ch} \frac{\pi y}{b} \cos \frac{\pi x}{a} + \frac{1}{2} \left(P_1 y^2 + P_2 x^2 \right) + F(x, \; y), \tag{19}$$

164

where C_1, D_1, γ_1, δ_1, P_1, and P_2 are constant quantities, and F (x,y) is a special solution of the compatibility equation for the deformations.

M. S. Ganeeva has analyzed the results obtained for the case of a square curvilinear panel with the parameters a = b = 20 cm, h = 0.1 cm, and L = $12.4 \cdot 10^{-6}$ 1/°C. The solution was found for a plate with different radii of curvature ($K_R = a^2/Rh$ is the curvature parameter), and gave the following results:

K	0	1.22	2.35	3.51	4.68	5.84	7.01	8.17	9.34
$(t_0)^0_{cr}$	1.7	20.3	67	146	257	399	573	779	1015

From the solution found (see also Fig. 5), the author draws the conclusion that there is an increase in the critical temperature at which loss of stability begins, if the curvature parameter K_R is increased.

The problem of the stability of a curvilinear panel ceases to have meaning for definite values of K_R, since the critical temperature approaches or exceeds the melting point of the material.

The approximate solution found for the problem gives values of the critical temperature that are somewhat too high, but there is only a small amount of difference between the results for the first and the second approximation as found by M. S. Ganeeva.

Some problems in the stability of circular plates under nonuniform heating have been solved by É. I. Grigolyuk [4].

Consider (see the paper by Grigolyuk [4]), a circular plate of constant thickness h, symmetrically heated about the axis according to some arbitrary law along the radius. We shall assume the temperature constant over the thickness of the plate. Here, the heating temperature T at any point of the plate will be a function solely of the distance r from the center (Fig. 6).

The material of the plate is assumed to be elastic, and the physical parameters of the material, E, ν, and α are independent of temperature.

The problem of finding the critical temperatures in this case reduces to solving the system of differential equations

$$\frac{\partial^2 u}{\partial \rho^2} + \frac{1}{\rho} \frac{\partial u}{\partial \rho} - \frac{u}{\rho^2} = (1 + \nu) b\alpha \frac{dT}{d\rho},$$ (20)

$$\frac{\partial^2 v}{\partial \rho^2} + \frac{1}{\rho} \frac{\partial v}{\partial \rho} - \left(1 + \frac{N_1 b^2}{D} \rho^2\right) \frac{v}{\rho^2} = 0,$$

where u is the radial displacement of the points in the median surface, v is the angle of inclination of the normal to the median surface after deformation, ρ = r/b is the relative radius at any point, and N_1 is the radial normal force,

$$N_1 = \frac{Eh}{(1 - \nu^2) b} \left[\frac{\partial u}{\partial \rho} + \frac{\nu}{\rho} u - (1 + \nu) b\alpha T\right].$$ (21

The temperature field is assumed to be given by the following power law:

$$T = t_0 + t_1 (1 - \rho)^n,$$ (22)

where n is an arbitrary exponent.

In solving this problem, the author uses Galerkin's method, giving the expression for the angle of inclination of the normal to the median surface in the form

$$v = \frac{C_1}{\rho} + C_2 \rho + C_3 \rho^3,$$ (23)

165

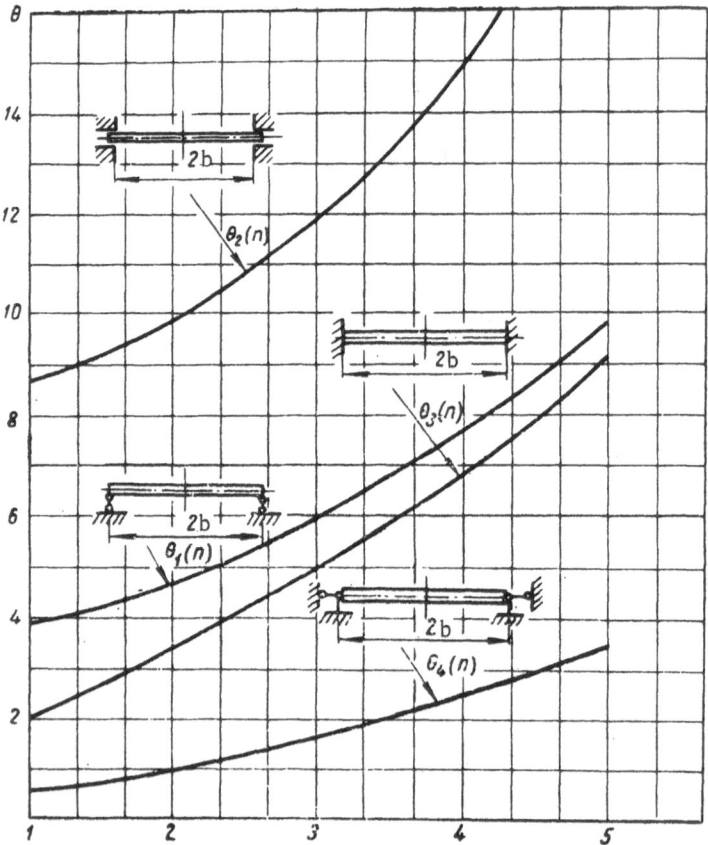

Fig. 7. Critical temperature parameter of a solid circular plate
for different conditions of support.

where C_1, C_2, and C_3 are some parameters, two of which may be expressed in terms of the third.

For a solid circular plate freely supported at the edge at r = b Galerkin's method gives the following expression for the critical temperature:

$$(t_1)_{cr} = \theta_1(n) \frac{\delta^2}{\alpha} , \qquad (24)$$

where δ = h/b is the relative thickness of the plate, and $\theta_1 = \theta_1(n)$ is a numerical coefficient.

Calculating $\theta_1(n)$ for $\nu = \frac{1}{3}$ for different values of n, gives the following results:

n	1	2	3	4	5
$\theta_1(n)$	3.83	4.60	5.98	7.71	9.73

For a plate clamped around the periphery, the critical temperatures may be calculated from the formula

$$(t_1)_{cr} = \theta_2(n) \frac{\delta^2}{\alpha} , \qquad (25)$$

$$\theta_2 = \theta_2(n):$$

n	1	2	3	4	5
$\Theta_2(n)$	8.69	9.69	12.0	15.0	18.4

The coefficients $\Theta(n)$ are given in Fig. 7 for other boundary conditions.

The paper by Grigolyuk [4] also discusses the stability of uniformly heated flat rings for two different boundary conditions.

For a ring with the inner edge free and the outer edge rigidly clamped, the critical temperatures may be found as

$$(t_0)_{\mathrm{cr}} = \lambda_3(\beta)\frac{\hat{\delta}^2}{\alpha}. \tag{26}$$

For a free inner and a freely supported outer edge

$$(t_0)_{\mathrm{cr}} = \lambda_4(\beta)\frac{\hat{\delta}^2}{\alpha}, \tag{27}$$

where $\beta = a/b$ is the ratio of the radii, and λ_3 and λ_4 are numerical coefficients found from Fig. 8.

Buckling of a Cylindrical Shell under Heating

The stability of cylindrical shells under heating is discussed principally in the papers by Hoff [1] and Zuk [5].

Hoff [1] comes to the conclusion that buckling of a cylindrical shell is most likely in the case where the temperature field is a function of the peripheral or axial coordinate. If the temperature only changes in the radial direction, no large deformations are produced on heating.

If the temperature only varies in the peripheral direction, the result of the temperature stresses is similar to what occurs in flat plates with the same temperature distribution over the width. For this reason, there is danger of buckling, although it is not as large as in a flat plate, since the curvature of the shell produces a clearly defined stabilizing effect.

For the problem of stability under nonuniform heating in the axial direction of the shell, Hoff draws an analogy with a cylinder loaded by variable annular compression which changes along the axis according to the law

$$\sigma_\varphi = RE\sum_{m=0}^{\infty} s_m \cos\frac{m\pi x}{\lambda}, \tag{28}$$

where R is a coefficient, E is the modulus of elasticity, λ is equal to L/a, and L and a are the length and radius of the shell, respectively.

The annular compression stresses may be regarded as the result of uniform heating combined with clamping the shell at the ends (Fig. 9).

Using the Donnel equation, the author in this case finds the differential equation of the bending in the form

$$\left.\begin{aligned}
\nabla^8 w + 4K^4\frac{\partial^4 w}{\partial x^4} + 4K^4\nabla^4\left(\frac{\sigma_\varphi}{E}\frac{\partial^2 w}{\partial\varphi^2}\right) &= 0, \\
\nabla^4 u &= \nu\frac{\partial^3 w}{\partial x^3} - \frac{\partial^3 w}{\partial x\partial\varphi^2}, \\
\nabla^4 v &= (2+\nu)\frac{\partial^3 w}{\partial x^2\partial\varphi} + \frac{\partial^3 w}{\partial\varphi^3},
\end{aligned}\right\} \tag{29}$$

where $4K^4 = 12(1-\nu^2)(a/h)^2$, and x is the dimensionless axial coordinate.

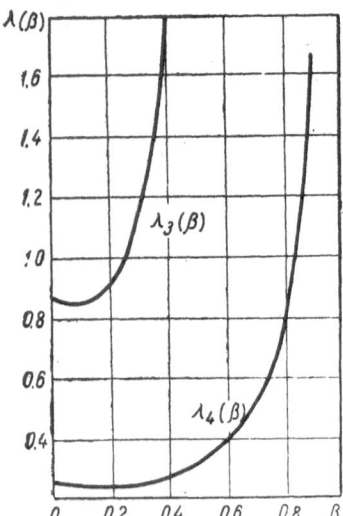

Fig. 8. Critical temperature parameter of a flat circular ring for different boundary conditions.

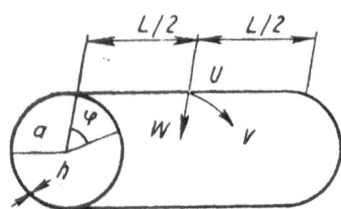

Fig. 9. Notation used for circular cylindrical shell.

In looking for a solution in the form of trigonometric series

$$
\left.
\begin{aligned}
w &= \cos n\varphi \sum_{m=1}^{\infty} a_m \sin \frac{m\pi x}{\lambda}, \\
u &= \cos n\varphi \sum_{m=1}^{\infty} b_m \cos \frac{m\pi x}{\lambda}, \\
v &= \sin n\varphi \sum_{m=1}^{\infty} c_m \sin \frac{m\pi x}{\lambda},
\end{aligned}
\right\} \tag{30}
$$

Hoff obtains the characteristic equations as an infinite series of linear equations, and limits himself to an analysis of the results obtained.

For uniformly distributed annular compressive stresses, the critical stresses are of the form

$$
(\sigma_0)_{cr} = 0.92\, E \left(\frac{h^3}{aL^2} \right)^{1/2}. \tag{31}
$$

In analyzing any possible departure from uniform compression, the author comes to the conclusion that buckling is less probable for high peripheral compressive stresses around the ends of the shell. The danger of buckling becomes much greater if the compressive stresses are concentrated in the middle of the cylinder.

The similar problem of thermal buckling of a cylindrical shell clamped at the edges has also been discussed by Zuk [5].

In contrast with Hoff's work, Zuk considers the less general case of annular stresses, assuming

$$
\sigma_\varphi = \frac{E\alpha T}{2} \left[1 + \cos\left(\frac{2\pi x}{L} \right) \right]. \tag{32}
$$

Using the Galerkin method as applied to the Donnel equation, and assuming

$$
w = \sin \frac{\pi y}{\lambda} \sum_{m=1,3,\ldots}^{\infty} a_m \sin \frac{\pi x}{L} \sin \frac{m\pi x}{L} = \sum_{m=1,3}^{\infty} a_m \Phi_m,
$$

Zuk found a simpler solution of the problem.

The final value of the critical temperature for a shell clamped at the ends and free with respect to axial displacements is given by Zuk in the form

$$
T_{cr} = \frac{\pi^4 r^2 t^2 (6L^8 + 32L^6\lambda^2) + 384L^4\lambda^8}{12(1 - \nu^2)\, \pi^2 a r^3 L^4 \lambda^2 (7L^4 - L^2\lambda^2)}. \tag{33}
$$

The author recommends that the numerical values of the critical temperatures be found graphically, using a series of values of the parameter λ and finding the minimum in the $T_{cr} = f(\lambda)$ curve for the given dimensions of the shell and the given values of E, α, and ν.

LITERATURE CITED

1. Hoff, N., "Buckling at high temperature," J. Roy. Aeronaut. Soc., No. 563 (1957).
2. Klosner, I., and Forray, M., "Buckling of simply supported plates under arbitrary symmetrical temperature distributions," IAS, No. 3 (1958).

3. Ganeeva, M. S., "Stability of a rectangular cylindrical panel rigidly clamped at the edges located in a nonuniform temperature field," Uch. Zap. Kazanskogo Univ. 116, No. 1 (1956).

4. Grigolyuk, É. I., "Some problems in the stability of circular plates under nonuniform heating," Inzh. Sb., Vol. 6 (1950).

5. Zuk, W., "Thermal buckling of clamped cylindrical shells," IAS, No. 5 (1957).